T0342432

Intrapreneurship Management

IEEE Press
445 Hoes Lane
Piscataway, NJ 08854

Intrapreneurship Management

Concepts, Methods, and Software for Managing Technological Innovation in Organizations

Rainer Hasenauer
Institute of Marketing Management
Vienna University of Economics and Business, Austria

Oliver Yu
Lucas College of Business, San Jose State University
San Jose, CA, USA

Published by John Wiley & Sons, Inc., Hoboken, New Jersey.
Published simultaneously in Canada.

For general information on our other products and services or for technical support, please contact our Customer Care Department within the United States at (800) 762-2974, outside the United States at (317) 572-3993 or fax (317) 572-4002.

Wiley also publishes its books in a variety of electronic formats. Some content that appears in print may not be available in electronic formats. For more information about Wiley products, visit our web site at www.wiley.com.

Library of Congress Cataloging-in-Publication Data

Names: Hasenauer, Rainer, author. | Yu, Oliver S., author.
Title: Intrapreneurship management : concepts, methods, and software for managing technological innovation in organizations / Rainer Hasenauer, University of Economics and Business, Institute of Marketing Management, Austria, Oliver Yu, Lucas College of Business, San Jose State University, San Jose, California, USA.
Description: Hoboken, New Jersey : Wiley, [2024] | Includes bibliographical references and index.
Identifiers: LCCN 2024013139 (print) | LCCN 2024013140 (ebook) | ISBN 9781119837725 (hardback) | ISBN 9781119837732 (adobe pdf) | ISBN 9781119837749 (epub)
Subjects: LCSH: Technological innovations–Management. | Diffusion of innovations–Management.
Classification: LCC HD45 .H34345 2024 (print) | LCC HD45 (ebook) | DDC 658.5/14–dc23/eng/20240419
LC record available at https://lccn.loc.gov/2024013139
LC ebook record available at https://lccn.loc.gov/2024013140

Cover Design: Wiley
Cover Image: © SEAN GLADWELL/Getty Images

Set in 9.5/12.5pt STIXTwoText by Straive, Pondicherry, India

We dedicate this book to our better halves, Gudrun and Evelyn, who not only endured but also heartily supported the long development of the book.

Rainer and Oliver

Contents

About the Authors

Rainer Hasenauer is Honorary Professor of Marketing and Lecturer in Marketing of High-Tech Innovation and Technology Marketing at the Institute of Marketing Management, Vienna University of Economics and Business (WU Vienna). He is an entrepreneur, a co-founder, and a business angel. He is also a business developer focusing on innovative technologies. His primary teaching and research interests lie in market entry of high-tech innovation in B2B markets and in measuring Innovation Half-Life and technology acceptance in B2B markets. His research work is predominantly project-driven for B2B markets and comprises community-based innovation, marketing testbeds for market entry, and multidisciplinary communication in high-tech innovation. Applications include satellite navigation and remote sensing, robotics, sensors, functional materials, flow batteries, and remote power supply. He serves on the advisory boards of high-tech investment groups, high-tech start-up incubators, and the supervisory board of a global market leader in the field of safety-critical real-time communication systems.

Dr. Oliver Yu is Executive in Residence and Faculty Emeritus at San Jose State University (SJSU); Founder of the STARS Group, a technology strategy consulting firm; Co-founder of GA2000 LLP, a startup for the design and operation of innovation parks, with its first 20-acre innovation park in Bengbu, China; and Co-founder of International Society of Innovation Methods (ISIM), a professional society for promoting scientific approach to creative problem-solving. Prior to these, he was for 11 years Director of Energy and Technology Strategies at Stanford Research Institute (SRI) assisting corporations and governments around the world, and

15 years Manager of Planning Analysis at Electric Power Research Institute (EPRI), responsible for global energy analysis and EPRI-wide research planning. Dr. Yu has published over 80 professional papers and authored and co-authored 6 technical books, including *Technology Portfolio Planning and Management*, Springer; *Technology Management and Forecasting*, Tsinghua University Press; and *Advances in Technology and Innovation Management*, IEEE. He has been a consulting associate professor at Stanford and an invited lecturer at numerous research institutions, including Visiting Professor of Business Innovations at National Kyushu University in Japan, special lecturer at the Summer School for Global Entrepreneurship at the University of Muenster in Germany, twice a special lecturer at Chinese Academy of Science, and twice a keynote speaker at both the Asia-Pacific Technology Foresight Conferences and the Portland International Conferences for Management of Engineering and Technology (PICMET), as well as at the INOTEX international innovation conference in Iran. Dr. Yu has also organized many technical conferences, including General Chair of both a national and the first international meeting of the Institute for Operations Research and Management Sciences (INFORMS), a Global Innovation Forum jointly sponsored by IBM, SRI, and SJSU, and the International Symposium of Innovation and Entrepreneurship of the IEEE Technology and Engineering Management Society (TEMS), for which he was a board member. Dr. Yu is a senior member of IEEE and a fellow of PICMET.

Preface

Since being coined in the 1970s, Intrapreneurship, or Internal Innovation in an organization, has gained increasing recognition, with the awareness that most people work for organizations, and internal innovations in organizations have been responsible for the majority of Innovations in both technology and management methods. However, there have not been many practical books on intrapreneurship development and management. Through a series of professional and personal meetings over the years, we both recognize the opportunity to use our extensive experience and knowledge in the development and management as well as the various readiness assessments of internal innovation projects to make a useful contribution to fill this important void. With the support of IEEE Press, especially through Gerald (Gus) Gaynor, Vice President of Publication of IEEE's Technology and Engineering Management Society, and Dr. Tariq Samad, Founding Editor of Wiley/IEEE Press Book Series on Technology Management, Innovation, and Leadership, we received permission to develop this book with Wiley Science-IEEE Press. With more than 150 video and personal meetings over the span of 3 years and the generous support of many contributors, we have finally completed the book. By integrating diverse concepts on understanding human nature and organizational culture and combining them with practical methods for creative problem solving and productive team building, as well as a proven Intrapreneurship readiness assessment methodology and software, we hope that this book will provide not only theoretical perspectives but also useful tools to both an organization and an aspiring Intrapreneur as well as academicians and students interested in intrapreneurship development and management.

Rainer Hasenauer,
Lindengasse 2/9, A-1070 Wien, +436641609106,
rainer.hasenauer@wu.ac.at.

Oliver Yu,
630 Woodside Drive, Woodside, CA, USA 94062, 1+650-922-0882,
oliver.yu@sjsu.edu.

Acknowledgments

We would like to especially acknowledge the following seasoned experts for their special contributions to the book.

Christian Rathgeber, Project Manager and Product Owner for the READINESSnavigator© software and AI Specialist at ONTEC AG in Vienna with over 40 years of experience as an IT developer and manager, is the key contributor to the Intrapreneuship READINESS navigator© software development for the book.

Matthew Schlegel, Principal of Schlegel Consulting (www.evolutionaryteams.com) and a best-selling author committed to developing highly effective, style-diverse teams, is a major contributor to the applications of the Enneagram to productive Innovation Team building and creative problem-solving for the book.

D. Daniel Sheu, Ph.D., Professor Emeritus for Industrial Engineering at National Tsing-Hua University in Taiwan, founder of the International Society of Innovation Methods, and a renowned expert on TRIZ (Russian acronym for the Theory of Inventive Problem-Solving), has authored 13 books and over 200 professional papers on TRIZ and other innovation methods and applications and conducted TRIZ and other innovation method training around the world. He is a major contributor to the applications of TRIZ to creative problem-solving for the book.

Michael Gold, a specialist in consumer psychological assessment and an intrapreneur during a 30-year career at the former Stanford Research Institute (now SRI International), following 8 years of development-engineering experience in Silicon Valley, has made valuable contributions to the book in reviewing its contents on Intrapreneurship, human needs and wants, and providing information on the cultural history of SRI.

1

Introduction and Overview

The purpose of this book is to provide practical concepts, methods, tools, and software for developing and managing Internal Innovators, or Intrapreneurs, in organizations, as well as for academicians and students of Intrapreneurship Management. In this introductory chapter, we will first develop a simple yet comprehensive definition of innovation that emphasizes implementation and impact. We will then discuss the basic differences between the Innovator who develops and implements an innovation as an Entrepreneur and an Intrapreneur, or Internal Entrepreneur in an existing organization, and the special importance of Intrapreneurship and its management not only for organizations but also for society as a whole in the 21st century, which is the basic motivation for this book. We will next summarize the focus, approaches, emphases, and major contributions of the book to the study of Intrapreneurship Management. Finally, we will present an overview of the organization of the remaining chapters of the book.

1.1 Innovation: A Simple Yet Comprehensive Definition

Innovation is a term with many definitions. For example, the Merriam-Webster dictionary [1] defines innovation as a new idea, method, or device. Another authoritative source, the Organization for Economic and Cooperation and Development (OECD), provides a more detailed definition [2]: "An innovation is the implementation of a new or significantly improved product (good or service), or process, a new marketing method, or a new organizational method in business practices, workplace organization or external relations."

However, both of these widely accepted definitions have their limitations. The dictionary definition fails to distinguish innovation from invention, which is also a

Intrapreneurship Management: Concepts, Methods, and Software for Managing Technological Innovation in Organizations, First Edition. Rainer Hasenauer and Oliver Yu.

Published 2024 by John Wiley & Sons, Inc.

new idea, method, or device. On the other hand, the OECD definition is too restrictive, as innovation is more than just products, processes, or methods. There are important theoretical innovations such as Newtonian laws and evolutionary theory that have greatly improved our understanding and management of the physical environment. There are also philosophical and ideological innovations like Confucianism, democracy, and Marxism that have fundamentally influenced our sociopolitical systems. Furthermore, there are innovations in literature and the arts that have significantly broadened our enjoyment and appreciation of creativity.

Given the limitations of traditional definitions, we offer a simple yet comprehensive definition:

Innovation is an idea implemented with significant impact.

This definition emphasizes that an idea, whether new or existing, is the starting point for innovation. However, an idea alone is not enough to qualify as innovation. The idea must be implemented in a way that achieves significant impact through widespread and/or long-lasting acceptance by final adopters, which is often a complex and challenging process. The emphasis on implementation and impact is the key differentiator between our definition and conventional ones. It also highlights a significant challenge of innovation: to achieve significant impact, large investments of effort are often required to understand market needs, build team, seek support, develop product, and attract adopters, while these investments carry major risks that they may not yield the desired results. This challenge is often recognized but not fully appreciated by many potential innovators, leading to discouraging early failures. The central objective of innovation management is to effectively address this challenge.

Moreover, the idea to be implemented need not be limited to new or improved products, processes, or methods, although they are the main focus of this book. It can also be a new concept or a newly uncovered concept for the innovator, organization, or adopter. For instance, the integration of economic benefit with ecologic and societal benefits as the value framework adopted in this book is an innovation that recognizes new values for all innovation participants. The development and expanding adoption of Environmental, Social, and Governance (ESG) scores for investment strategy is one form of implementing this conceptual innovation.

Innovations may be implemented through various means including market mechanisms and voluntary choices by individual adopters, as well as mandatory government regulations, such as automobile emission control. The differences in implementation can affect the effective management of the innovation process. Nevertheless, in a democracy, government regulations may be viewed as collective choices made through a political market mechanism.

Finally, the effectiveness of an innovation is measured by its impact, which varies in magnitude and duration. There are many measures of magnitude, including the number of adopters, the integrated value of the innovation, and other tangible or intangible benefits to the Innovator or the adopter. The duration of an innovation measures whether its impact is transitory or long-lasting. The total impact or effectiveness of an innovation can be measured by integrating its magnitude over time. As impact can be tangible or intangible and innovations are generally evolving and cumulative, quantifying the impact of a single innovation is an important yet often difficult task.

1.2 Special Importance of Intrapreneurship Management

It is widely acknowledged that innovation has been the driving force behind human civilization. Following major breakthroughs in information, communications, and biomedical technologies, along with the rapidly rising global trend of high-tech startups, innovations, particularly those in technology and management methods, have generated unprecedented interest worldwide since the 1970s. Today, practically all major organizations, including governments, corporations, and educational institutions, are actively promoting the importance of innovation, motivated by its much-observed impacts on economic productivity and competitiveness, as well as employment opportunities. There is a popular belief that future innovations, if well managed, can provide solutions not only to many economic problems but also to sociopolitical difficulties around the world.

However, despite the attention and glamour, innovators as Entrepreneurs who develop and implement innovations through the founding of independent organizations are still a small minority of the workforce, mainly because of their necessary yet uncommon personality traits of extraordinary risk-taking, determination, and persistence.

On the other hand, most people work for existing organizations, including those rising from startups. At the same time, all organizations need innovation to survive and thrive. By sheer numbers, most innovations have been and will continue to be developed by innovators within organizations, or Intrapreneurs. In fact, an in-depth analysis of the top 30 innovations between 1989 and 2009 [3] shows that, although many Entrepreneurs provided seminal ideas, out of the 30 top innovations, 22 were initially conceived and 23 were fully developed by Intrapreneurs in existing organizations, and 16 achieved full impact through Intrapreneurs in competitor organizations.

Nevertheless, because Intrapreneurs are largely unsung heroes, their contributions have not attracted widespread public attention. Moreover, different from the intrinsically self-starting Entrepreneurs, Intrapreneurs generally need to be developed and managed in an organization. This development and management process is highly interactive and complex between the managers and employees of the organization. As a result, available business and technical literature lacks comprehensive and in-depth studies of the important but seemingly less tractable subject of how to effectively develop and manage Intrapreneurs or Intrapreneurship Management in an organization. Yet the importance of Intrapreneurship Management has become even more critical given the unprecedented changes in the 21st century in socioeconomic conditions, human aspirations, technological growth, and the physical environment, which pose major challenges and demand new thinking about innovation management in an organization. This special importance of effective Intrapreneurship Management has been the basic motivation for this book.

1.3 Focus, Approaches, Emphases, Theme, and Major Contributions of the Book

Focusing on developing and managing Intrapreneurs in 21st century organizations, this book primarily addresses innovations in technology, products (goods or services), and management methods that are implemented through market mechanisms based on individual choices or through government policies and regulations, with generally accepted measures of economic, ecologic, and societal impacts.

The book uses an integrated systems analysis approach to study the innovation process, especially in an organization, and develops a joint investment perspective to balance and align the values and risks of each innovation participant, including Intrapreneur, Internal Supporter, and Final Adopter, for investing time, effort, financial and other resources to develop and implement an innovative product for significant impact. The book further develops a new perspective on human needs and wants as the basis for assessing these investment values and risks through an integrated economic–ecological–equity (EEE) value framework.

For effective Intrapreneurship Management, the book focuses on the importance of marketing, which includes mutual marketing between Intrapreneur and Internal Supporter in an organization. The book develops a unified approach to marketing for motivating and committing Intrapreneurs, Internal Supporters, and Final Adopters to their joint investment in innovation development and implementation. Effective marketing forms the foundation for an innovation culture within an organization and provides the basis for widespread innovation adoption in the face of many external challenges. The book then presents major approaches to creative idea generation and problem-solving for innovation development in a productive organizational culture with complete understanding of the needs of the Final Adopter. Marketing and innovation are, therefore, the central themes of this book, in line with Peter Drucker's observation that "The business enterprise has two, and only two, basic functions: marketing and innovation." [4]

Since management is still largely an art, and creativity is far from a science, the emphasis of the book is on providing practical concepts and proven tools for effective innovation management and creative problem-solving for organizations. The book emphasizes not only new concepts, perspectives, and approaches but also in-depth analysis, replicable successful case studies from both global corporations and small to medium technology companies, useful discussion topics, and training exercises. It is important to note that most of the concepts introduced are qualitative, and the tools presented are mostly process-oriented, highly interactive, and experiential. As a result, expert-facilitated training exercises and continuous practices are essential to the successful learning and applications of the information provided in this book.

Combining the integrated system analysis approach with their extensive experiences as successful managers of innovations and experienced researchers of market behaviors, business psychology, and creativity methods, the authors methodically identify three key elements for effective Intrapreneurship Management: Organization, Market, and Technology Readiness. Studies of these key elements are then used to generate a set of questionnaires and checklists as the basis for assessing the overall Intrapreneurship Readiness of an organization and a specific innovation project.

These questionnaires and checklists have been used to extend a family of proven Innovation Readiness assessment software tools to the development of a module-based Intrapreneurship READINESS navigator (IRN©) software for Intrapreneurship Management. The book concludes with detailed descriptions of the IRN© software architecture and generic application examples. Again, this IRN© software is a living, learning tool that can be customized and adapted to the applications of different types of organizational structures and innovation projects.

Major contributions of the book include:

- A simple but comprehensive definition of innovation that emphasizes implementation and impact, a systematic analysis of the innovation process, and a common investment perspective for different innovation participants to be used for aligning the balances of their individual values and risks in investing in innovation development and implementation.
- A systematic methodology for analyzing the process of Intrapreneurship Management through the assessment of the readiness of three key elements: Organization, Market, and Technology of Internal Innovation encompassing the organization, the Innovation Team, and the innovative project and product.
- A holistic understanding of human values based on fundamental needs and wants that applies not only to motivations of individuals but also to organizations and societies and culminates in an integrated EEE value framework for evaluating the Organization, Market, and Technology Readiness for Intrapreneurship Development and Management.

- A unified approach to marketing, including the mutual marketing between Intrapreneur and Internal Supporter in an organization, is the basis for motivating innovation participants to jointly invest their time, effort, financial, and other resources in the implementation of products (goods or services) developed from a creative idea to achieve major impact.
- A systematic identification and analysis of the individual sets of key elements for the Organization, Market, and Technology Readiness of Intrapreneurship Management, and a collection of major proven tools for organizational culture formation, Innovation Team building, creative problem-solving, innovative product development, and marketing.
- A set of interactive questionnaires and checklists generated from the key element analysis and a decision tree framework as the basis for constructing the IRN© software for Intrapreneurship Development and Management.
- A collection of case studies of successful innovative organizations and detailed descriptions and application examples of the IRN© software.

In summary, the book is both a research monograph and a practical reference for developing effective Intrapreneurs to increase success probabilities and potential impacts of innovations in 21st-century organizations. It applies an integrated system analysis approach to study the motivations of and interactions among the major participants of Internal Innovation in an organization: Intrapreneur, Internal Supporter, and Final Adopter, and develops practical concepts, tools, and software for assessing and increasing both the individual and the combined Organization, Market, and Technology Readiness for Intrapreneurship Development and Management. Finally, the book serves to help Intrapreneur, Internal Supporter, and Adopter to effectively communicate, collaborate, and implement innovations to fulfill increasingly higher levels of needs and wants to further the advances of human civilization.

1.4 Overview of Book Organization

Following the introduction and overview in this chapter, Chapter 2 of the book outlines the general structure of Intrapreneurship or the Internal Innovation process in an organization as a dynamic interactive system of key elements, including establishing innovative culture, understanding Adopter's needs, generating/adapting creative ideas, building an Innovation Team, developing/testing/improving prototypes, seeking/providing support, refining innovation, and developing organizational and other strategic and operational functions. It then uses an integrated systems analysis approach to identify the interactions among the key elements and the participants of the innovation process: Intrapreneur as

Innovation Team, Internal Supporter, and Adopter. The analysis focuses particularly on the special roles and interactions between the two principal participants of Internal Innovation, i.e., Intrapreneurship, in an organization: Intrapreneur and Internal Supporter, and identifies the three key elements for Intrapreneurship Management: Organization, Market, and Technology Readiness. Again, based on the integrated system's analysis approach, the chapter develops a common investment perspective among all innovation participants for maximizing their individual perceived expected values with an acceptable perceived expected risk for investing their time, money, effort, and other resources in implementing the products (goods or services) developed from a creative idea. The chapter highlights the importance of aligning the investment perspectives among all participants for effective innovation development and implementation. Furthermore, Chapter 2 studies the major changes in the business environment in the 21st century that affect the values of the participants and pose challenges to Intrapreneurship Management. It concludes with a set of generic challenges to Intrapreneurship Management and an integrated EEE value framework for assessing the readiness of Internal Innovation of an organization in the changing and challenging business environment.

Chapter 3 explores the foundation of Intrapreneurship Management, which is an understanding of the value and risk attitude of innovation participants based on human needs and wants in their investment decisions for implementing products (goods or services) developed from a creative idea. The chapter presents a new perspective that classifies and explores human needs and wants in the interactive physical–psychological and security–stimulation dual dimensions, which are further extended to the needs and wants of organizations and society as a whole. The chapter concludes with a discussion of the application of human needs and wants in the context of Intrapreneurship Management as the basis of an integrated EEE evaluation framework to be used in the IRN© software for assessing specific innovation projects in an organization.

Chapter 4 presents a unified marketing approach aimed at achieving the central goal of Intrapreneurship Management, which is to align the investment perspectives of all major innovation participants to effectively implement the products (goods or services) developed from a creative idea generated in an organization. The chapter describes the common stages of the marketing process and highlights the major barriers and potential solutions to effective marketing. While innovation marketing is mostly applied to the Final Adopter, the chapter shows that for Intrapreneurship Management, there is important mutual marketing between the Intrapreneur and Internal Supporter in an organization, and the unified approach can be applied to this internal mutual marketing to achieve the alignment of the perceived expected values and acceptable perceived expected risks of these two principal participants in an organization. By facilitating their joint investment in innovation implementation, this alignment forms the basis of an effective innovation culture in the organization. The chapter presents the key elements of Organization Readiness of Internal Innovation or Intrapreneurship: Innovative Culture, Internal Support, and Innovation Team. It first defines organizational culture for Intrapreneurship Management and emphasizes the importance of perceived behavior norms and codes of conduct. The chapter further emphasizes the importance of established internal support system for innovation and summarizes the successful practices in several major innovative organizations. It finally introduces two major proven approaches for effective team building and management: one based on the popular software development process of Agile and Scrum, and the other based on the ancient principle of Enneagram. Analysis of the key elements, the principles of team building approaches, and the integrated EEE value framework jointly form the basis for developing systematic questionnaires and checklists to assess the Organization Readiness level in Intrapreneurship Management.

Chapter 5 applies the unified marketing approach to develop the key elements of Market Readiness for a specific innovative product through a deep understanding of the needs and wants of the Final Adopter. The chapter first discusses the interaction between Market Readiness and Technology Readiness, the evolving marketing targets, advancing marketing tools, and major challenges to marketing to the Final Adopter. It then analyzes the key elements of Market Readiness and various approaches for exploring and responding to observable, hidden, and unknown needs and wants of the Adopter. Furthermore, the subject of nescience, or the science of exploring the unknown, will be discussed for market need assessment. These applications form the basis for assessing the Market Readiness level in the IRN© software.

Chapter 6 first studies the origin of psychological inertia as the mental block to creativity and the need for diversity as the antidote. It then presents various tools for creative idea generation and problem-solving, including brainstorming, design thinking, the ancient Enneagram, the traditional logical thinking approaches, and the systematic principles of TRIZ. The chapter further outlines the systematic process for innovative technology development from constructing proof of concept, conducting physical and economic feasibility tests, building prototypes, and refining technology. It then identifies and discusses the key elements of Technology Readiness and uses them as the basis for generating a set of questionnaires and checklists for assessing the Technology Readiness Level in the IRN© software.

Chapter 7 integrates the results of Chapters 4–6 as the basis for assessing the Organization Readiness for Intrapreneurship and the iterative Market and Technology Readiness of an innovation project. Due to the diversity and complexity of various types of organizations and different kinds of innovations, the chapter presents a general methodological approach for these readiness assessments through a set of sample questionnaires and checklists that have been successfully applied to a variety of organization and innovation projects in the past two decades. These sample checklists can be further expanded, modified, and customized by the assessment team for specific applications integrating their own capabilities and experiences. The methodological approach and the large set of sample checklists form the basis for the IRN© software in Chapter 8.

Chapter 8 presents a detailed outline of the IRN© software, which is a member of the READINESS navigator (RN©) family of software packages for innovation management. The IRN© software is specifically designed to support the dynamic interactive innovation system within an organization by providing criteria for the Organization, Market, and Technology Readiness levels, tailored to the individual roles of major participants in the Internal Innovation implementation process. It uses a synchronized and balanced approach to achieve stepwise progress toward the Organization, Market, and Technology Readiness levels. The software requires specifications of characteristics of the overall organizational culture, the internal support system, and the Innovation Team for assessing the Organization Readiness for Intrapreneurship. The software uses the integrated EEE value framework to assess the progress of an innovation project toward fulfilling all conditions for potential innovation to execute market entry and achieve its economic, ecologic, and social benefit potential. It specifically uses a large set of Market and Technology Readiness checklists and the associated decision rules for the iterative and interactive assessment of the readiness for an innovation project in its next steps, including readiness for commercialization, continued market research and technology development, or resource transfer and project abandonment. For an innovation project, the software represents the essence of a dynamic innovation culture by being flexible and agile in assessing readiness and risks of innovations developed in an organization. The chapter includes a detailed description of the input requirements and a full set of examples of the user interfaces for applications to both Organization Readiness assessment of an organization and the iterative Market and Technology Readiness assessment of an innovation project.

Finally, Chapter 9, the Epilogue, presents an overall summary of the book and an outlook on the future research and developments of Intrapreneurship Management.

There are two appendixes. Appendix A presents the broad studies and in-depth interviews of senior executives of three world-renowned, long-term innovative companies: 3M, IBM, and SRI International. It summarizes the innovative culture, innovation achievements, and successful Intrapreneurship experiences and provides valuable insights into Intrapreneurship Development and Management from these world-class organizations. Appendix B presents a summary of the successful applications of the Market and Technology Readiness assessments to 26 innovation projects with detailed descriptions of the assessment process and the evolving Market and Technology Readiness trajectory of five selected cases conducted over the last 15 years. It provides a set of successful application experiences for the assessment methodology and software developed in the book.

Summary

In this first chapter, we have developed a simple yet comprehensive definition of innovation that emphasizes implementation and impact. We have also discussed the basic differences between the Innovator who develops and implements an innovation as an Entrepreneur and as an Intrapreneur, or Internal Entrepreneur in an existing organization, and the special importance of Intrapreneurship and its management not only for organizations but also for society as a whole in the 21st century. We have further summarized the focus, approaches, emphases, and major contributions of the book to the study of Intrapreneurship Management, and ended this initial chapter with an overview of the organization of the remaining chapters of the book.

Glossary

Entrepreneur: An innovator who seeks external support and forms an independent organization to develop and implement an innovation product.

Innovation: Simply an idea implemented with significant impact.

Innovation development and implementation: Innovative idea development and product implementation with special emphasis on implementation.

Innovation product: A product that is either a good or a service developed from a creative idea.

Innovation project: The project for developing and implementing an innovative product. Notice that a product based on a creative idea is innovative, while a project focuses on development and implementation.

Innovation Team: Synonymous to Innovator but emphasizing the team nature.

Innovator: The innovation participant who develops product from a creative idea; it is an Intrapreneur in an organization and also by nature an Innovation Team.

Intrapreneur/Internal Innovator: An innovator inside an organization, which is by nature an Innovation Team.

Intrapreneurship: Internal innovation in an organization.

Intrapreneurship Management: Development and management of innovation and Intrapreneur in an organization.

Discussions

- Discuss the sufficiency of the simple definition of innovation and possible modifications.
- Amplify the importance of Internal Innovation in an organization and identify the major Internal Innovations you may know of.
- Further research the special personality traits of an Entrepreneur and compare them to those of an Intrapreneur.
- Review the complexity of Intrapreneurship Management.
- Critique the approaches used in this book for the study of Intrapreneurship Management.

References

1 (2023). Merriam-Webster Dictionary.
2 OECD (2005). *The Measurement of Scientific and Technological Activities: Guidelines for Collecting and Interpreting Innovation Data: Oslo Manual*, 3e. OECD.
3 Krippendorf, K. (2019). *Driving Innovation from Within*. Columbia Business School Publishing.
4 Drucker, P. (1954). *Practice of Management*. Harper & Row.

2

The Internal Innovation or Intrapreneurship Process

In this chapter, we will first describe the Internal Innovation or Intrapreneurship process as an interaction system of key elements and identify the major participants. We will then present a common investment perspective for all participants and discuss the generic and special challenges to Intrapreneurship in the 21st century.

2.1 The Internal Innovation Process as an Interactive System

Based on the simple definition given in Chapter 1, innovation is not a static idea, but a dynamic process that, specifically for internal innovation in an organization, involves the development and implementation of a creative idea in an organization through an interactive and iterative system of key elements as shown in Figure 2.1.

This process engages several major participants and various types of resources over time. It begins with an innovative organizational culture that motivates the identification of needs, which include not only those of the Adopter but also the organization itself. The culture would also promote creative idea generation or adaptation in response to explicit or implicit internal and/or external needs/wants. The Innovative Culture also encourages the formation of an Innovation Team to develop a proof of concept of the innovative product and provide initial support, both monetary and nonmonetary. Initial prototype, minimum viable product, and trial rollout are then carried out with additional support. The next steps include refinement and rollout of innovative products, expansion of

Intrapreneurship Management: Concepts, Methods, and Software for Managing Technological Innovation in Organizations, First Edition. Rainer Hasenauer and Oliver Yu.

Published 2024 by John Wiley & Sons, Inc.

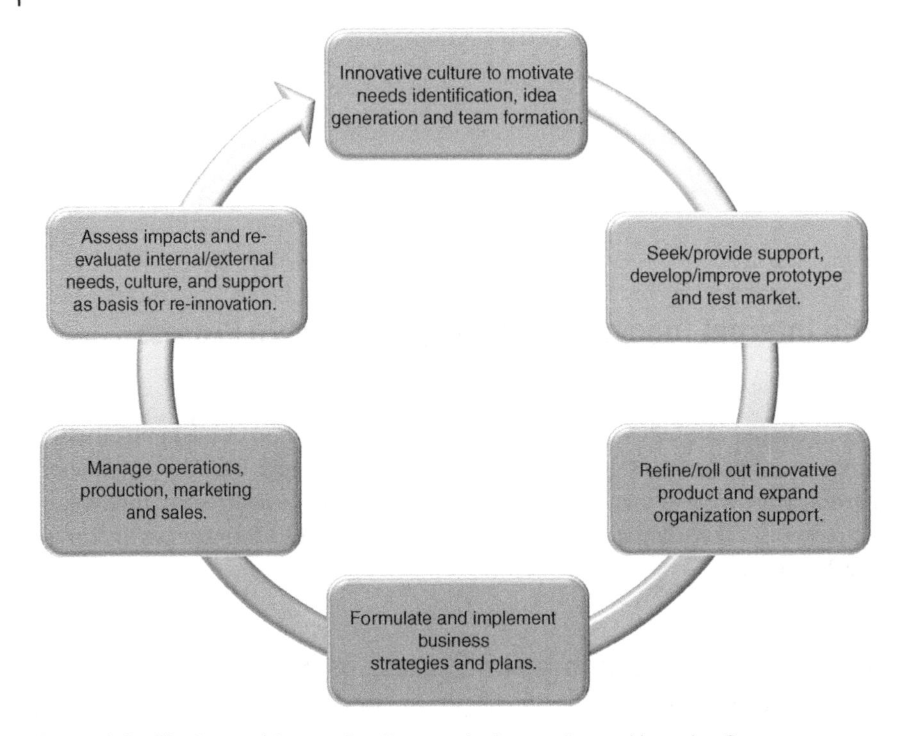

Figure 2.1 The Internal Innovation Process: An Interactive and Iterative System

organizational support, formulation and implementation of business strategies and plans, and management of operation, production, marketing, and sales, which are common to all business development and operation. Finally, there will be an assessment of the impact of the innovation development and implementation as well as a re-evaluation of the internal/external needs/wants, the organizational culture, and support to improve the idea generation, team formation, and innovative product development and implementation for re-innovation.

It is important to note that Figure 2.1 is a simplified depiction of the innovation process. Generally, innovation is cumulative and iterative. Thus, there may be many iterations between the first two key elements, unique to new idea generation and implementation, before going through the other operation-oriented gray-colored key elements. Furthermore, this process can also be applied to innovations within each key element.

Since the innovation process is a dynamic interactive system of key elements, an integrated system analysis approach can be applied to study the effective management of Internal Innovation development and implementation.

This approach will study not only the characteristics of individual key elements but also their collective synergy and common operating principles in the process. These common principles can then be used as the basis for designing operating systems in an organization that will stimulate creative idea generation, strengthen implementation, and expand the impact of innovation. Specifically, the approach will first systematically identify the major participants of the Internal Innovation process, examine their roles and interactions in the key elements of the process, and study their individual and collective values for innovation implementation. It will then study the background environment of the innovation process and its relationships to the values of these participants and use these relationships to develop a unified framework for analyzing and optimizing the overall impact of the Internal Innovation process.

As contrast, the innovation process for an Entrepreneur generally does not start with the existing organizational culture. Such culture can provide great support to innovation if it is cultivating and effective but can be a major obstacle to innovation if it is bureaucratic and risk-avoiding.

2.2 Major Innovation Participants of Intrapreneurship

For all innovation, there are three major participants: Innovation Team, Supporter, and Adopter. For the Internal Innovation process, both the Innovation Team and Supporter are internal to the organization, and we will specifically label the Innovation Team as Intrapreneur and the Supporter as Internal Supporter. In contrast, for innovation process by Entrepreneurs, the Innovation Team will be Entrepreneur and the supporter will be external supporter, like venture capitalists and other investors.

We emphasize the Innovation Team rather than individual innovators because the innovation process involves many different functions performed by people with different capabilities, making it by nature a team effort. However, an effective team still must have an individual champion dedicated to the success of innovation implementation. The champion may, but not necessarily, be the idea generator/adapter and can sometimes even be an Internal Supporter.

Again, as contrast, the Innovation Team for an Entrepreneur generally forms an independent organization with external support and develops its own organizational structure, business strategies, production process, and marketing mechanism. As a result, due to the strong need to develop and manage a new organization, an Entrepreneur is generally more organization-oriented. On the other hand, working with an existing organization, Intrapreneur is mostly project-oriented. Moreover, Intrapreneur can take many different forms, from a team of independent contributors to a formal research group.

An interesting special organization that combines both Intrapreneurship and Entrepreneurship is the incubator, where the organization focuses solely on the internal development of Entrepreneurs, which will also be part of our study.

Supporters are the lifeblood of innovation. However, for Intrapreneurship, Internal Supporters are even more essential for innovation development and implementation in an organization. They provide not only direct financial support but also sponsorship, mentoring, general management, and other support to the Intrapreneur. The Internal Supporters can also include shareholders of the organization, who indirectly provide financial and other support to the Intrapreneur.

Additionally, for all innovation, supporters also include external suppliers, community supporters, and regulators. In particular, regulators, often regarded as obstructions to innovation, can actually be highly important to innovation implementation by providing community approval and acceptance as indirect support.

Lastly, the Adopter includes not just the end-user but also the distributor who chooses to represent the innovative product, facilitates its delivery, and often provides the needed training and maintenance and after-service to the end-user. In addition, Internal Innovation also includes Internal Adopter in the organization.

2.3 A Common Investment Perspective of All Innovation Participants

The integrated system analysis approach highlights an important common perspective among all innovation participants: they are making investment decisions in participating in the innovation implementation process. Participants invest their time, money, effort, and other resources in anticipation of the perceived expected net value (benefit–cost), and all these investments have inherently perceived expected risks of failing to achieve these values. This perspective may be obvious for Innovation Team and Supporter, but is also for Adopter in choosing to participate in the process by investing their time, money, effort, and other resources to *adopt* the innovation, which carries both the perceived expected values of excitement, convenience, or other benefits as well as the perceived expected risk that the innovation may be a financial and/or emotional loss for not being able to achieve the perceived expected net value.

This perspective has two major implications:

First, modern portfolio theory for investment analysis [1] can now be used to study the investment decisions of each participant in the innovation process. The theory shows that for each participant, there exists an envelope for all possible innovation investment portfolios, shown as the efficient frontier in Figure 2.2. An innovation investment portfolio on the efficient frontier has the highest perceived expected value for a given level of perceived expected risk, and the lowest

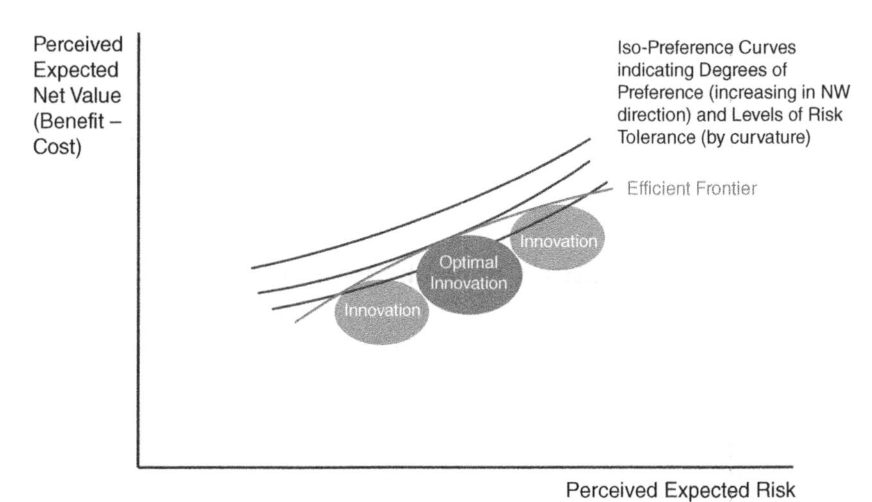

Figure 2.2 An Investment Perspective for a Participant in an Innovation Process

perceived expected risk for a given level of perceived expected value. Thus, the efficient frontier represents the collection of most desirable portfolios for each investor at different levels of perceived expected risk as shown by the three representative portfolios on the frontier. However, with varying levels of perceived expected risk, there is possibly an infinite number of desirable portfolios on the efficient frontier. To identify the best or optimal portfolio for a particular investor, one will need to first assess the investor's risk tolerance through the set of iso-preference curves in Figure 2.2, where all portfolios on an iso-preference curve are equally preferable to the investor to a particular degree. To avoid inconsistencies caused by one investment portfolio having different degrees of preference, the preference curves are necessarily parallel to one another, and the more northwest a curve lies, the greater the degree of the preference because of higher perceived expected values and lower perceived risks. Furthermore, a steep curvature of the iso-preference curve indicates a low-risk tolerance of the investor because, in this case, to achieve the same degree of preference for the investor, a small increase in perceived risk will need to be compensated by a proportionally large increase in perceived value. Modern portfolio theory also shows that the efficient frontier is always concave and the iso-preference curves are always convex in shape [2]. As a result, the best or optimal investment portfolio for a participant in the innovation process is the unique tangential point between the efficient frontier and the highest degree achievable iso-preference curve, as depicted in Figure 2.2.

Second, because each participant has its own unique optimal investment portfolio, it is critically important to align the investment perspectives of all major participants in the innovation process to achieve effective innovation management. The objective is to persuade and convince Innovation Team, Supporter, and Adopter to perceive that the particular innovation implementation is collectively the best investment of their individual required time, money, effort, and other resources.

2.4 Specific Applications of the Investment Perspective

Applying this investment perspective to different participants in the innovation process can yield interesting additional insights as shown in Figures 2.3–2.5.

Specifically, when applied to Innovation Team as shown in Figure 2.3, those with high-risk tolerance would generally view high-value and high-risk innovations as optimal investments, and they are generally the "Entrepreneurs." On the other hand, those with moderate risk tolerance would generally view medium-value and medium-risk innovations as optimal investments, and they are generally the "Intrapreneurs." Lastly, those with low-risk tolerance would generally view low-value and low-risk endeavors as optimal investments, and they may be called "nontrepreneurs." It is useful for all organizations to increase economic

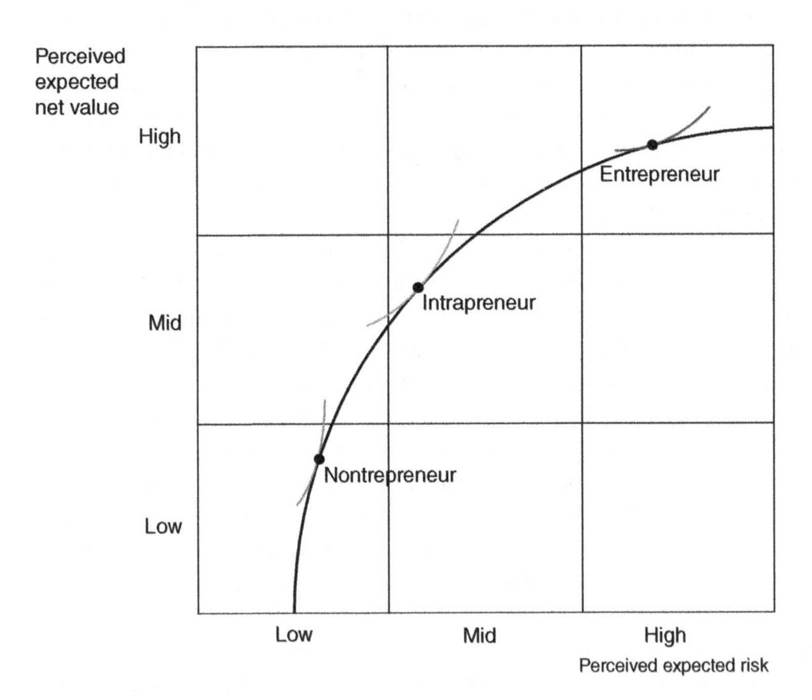

Figure 2.3 Investment Perspective for the Innovation Team

productivity and job satisfaction by finding effective ways to motivate these "nontrepreneurs" to become Intrapreneurs.

Similarly, the investment perspective can be applied to supporters in an innovation process as shown in Figure 2.4. In this case, supporters with high-risk tolerance, like advanced economies, leading corporations, and high-tech venture capitalists, generally view portfolios including high-value and high-risk projects, such as basic research, defense R&D, and new independent startup ventures as optimal investments. On the other hand, supporters with moderate risk tolerance, like newly industrialized economies, large corporations, and general business/ equity investors, tend to view medium-value and medium-risk portfolios, like applied research and corporate internal ventures, as optimal investments. Moreover, the moderate tolerance of risks and an organizational value perspective are also the key differentiators of Internal Supporter in an organization from external supporter such as venture capitalist. Last, supporters with low-risk tolerance, like developing economies, small/medium businesses, original equipment manufacturers, and fixed-income investors, generally view low-risk but low-value portfolios like incremental improvements as optimal investments. However, it may be useful and important for Innovators to find effective ways to motivate governments, businesses, and individual investors to increase their risk tolerance

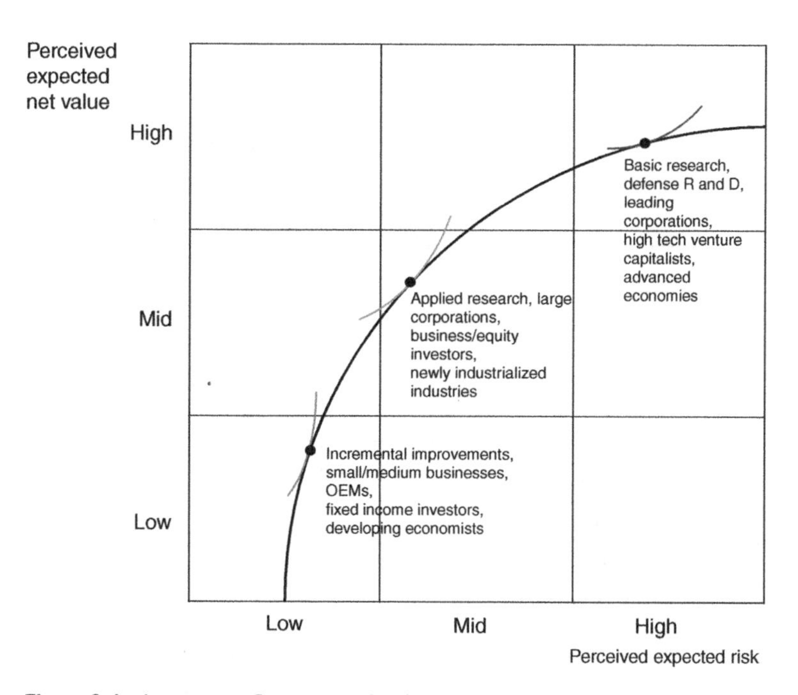

Figure 2.4 Investment Perspective for the Supporter

so that these innovation supporters can advance the economies, expand businesses, and develop investment acumens by making innovation investments with increasing gains in perceived values with increased but still tolerable perceived risks.

Finally, this perspective can also be applied to Innovation Adopters. As shown in Figure 2.5, those Adopters with high-risk tolerance generally view high-value and high-risk portfolios, like exciting or attractive but unproven new innovations, as optimal investments, and they are usually called "Early Adopters." On the other hand, Adopters with moderate risk tolerance tend to view medium-value and medium-risk portfolios, like proven innovations, as optimal investments, and they are often called "late followers." Lastly, Adopters with low-risk tolerance generally avoid new innovations because of the aversion of either risk or inconvenience in trying something new, and view low-value but low-risk portfolios with little innovations, as optimal investments of their time, money, effort, and other resources. They can be called "nonadopters." However, it is imperative for Innovators to find effective ways to increase the perceived values and reduce the perceived risks of their innovations so that the nonadopters can be motivated to be Adopters and the late followers can be motivated to be Early Adopters.

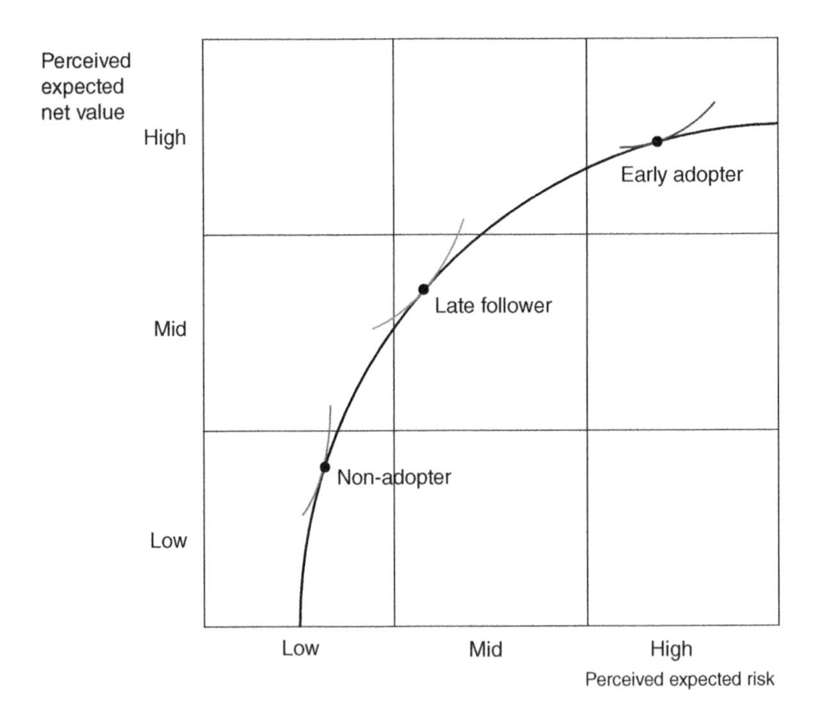

Figure 2.5 Investment Perspective for the Innovation Adopter

2.5 Alignment of Individual Investment Perspectives for Innovation Implementation

Because innovation implementation is a joint investment of the participants, to achieve impact, it is necessary to motivate and align *all* participants to perceive sufficient expected values at acceptable perceived risk. This point is especially important as organizations have often emphasized one participant or the other. Specifically, many organizations, due to the financial pressures of the market and indirect management control by the shareholders, have focused on maximizing immediate returns for their shareholders as Supporter, neglecting the long-term development of their organizations and employees. On the other hand, many other organizations have emphasized the importance of serving the needs/wants of the customers as Adopters and neglected the welfare of their employees. Then, there has been a recent trend for companies to emphasize the importance of satisfied employees as Innovation Team, recognizing that without motivated and productive employees there can be no innovation. The reality is that each emphasis focused only on one participant's values, and successful Intrapreneurship requires aligning the investment perspectives of all participants. To achieve this alignment, understanding the underlying motivations and values of the participants is critical, which is a major emphasis of the book.

Historically, the alignment of investment perspectives has been relatively simple for for-profit organizations, as both Supporter and Innovation Team share a primary interest in financial rewards for satisfying the needs of Adopter. However, the interest in financial rewards has declined in primacy, as businesses become increasingly concerned about and value environmental quality and social welfare [3]. Understanding these expanding values forms an integrated economic–ecologic–equity (EEE) value framework in this book for studying the joint investment perspective of the innovation participants in the changing and challenging background environment of the 21st century, which has also resulted in the analytical basis of the Intrapreneurship READINESS navigator (IRN©) assessment software.

Aligning investment perspectives is more complex for nonprofit organizations, such as governments. Democratic governments aiming at reducing the cost of regulations while promoting general welfare involve many conflicting public and private values of the innovation participants. To reduce complexity, this book mainly focuses on Intrapreneurship in for-profit organizations.

2.6 Generic Challenges to Intrapreneurship Caused by a Lack of Alignment of Investment Perspectives

Intrapreneurship, i.e., Internal Innovation in an organization, has the built-in advantages of readily available experienced mentorship, executive sponsorship, internal funding, organizational support, and other resources. However, in spite of these advantages, a lack of alignment of investment perspectives between Innovation Teams and Internal Supporter can generally create a set of generic challenges. These challenges include:

2.6.1 Risk-Averse Management Values and Organizational Culture

As an organization matures, the management tends to stick to proven successes and becomes less risk-taking than when the organization first started. The risk-averse management values will lead to an organizational culture that values stability and predictability, which, in turn, discourages investment in new ideas and may even lead to punishment of failures, which poses a major deterrent to Intrapreneurship. According to a study by McKinsey & Company [4], management value and organizational culture challenges are the most common barriers to successful Internal Innovation.

2.6.2 Resource Competition for the Internal Supporter

In existing organizations, most resources and employees are dedicated to the continued production and expansion of already successful products (goods or services). Implementing a new idea will create resource competition and may cause resistance from existing businesses, despite the Internal Supporter's interest in innovation. This lack of alignment of investment perspectives between Intrapreneur and Internal Supporter is a major cause of the "Innovator's dilemma" [5], while a study by the Global Innovation Index also notes that resource constraints are one of the most significant barriers to innovation in organizations [6].

2.6.3 Rigidity of Organizational Structure and Operating Procedures

As an organization grows, its internal structure and operating procedures become more formalized and hierarchical, leading to rigidity and limited autonomy to reduce administrative uncertainty and improve efficiency. However, the reduction in innovation value and resistance to innovation risks in the investment perspective of the organization caused by the rigid bureaucracy has been shown [7] to be a major challenge to Intrapreneurship as it restricts needed autonomy for creative idea generation, muddles communications for effective team formation, and creates organizational inflexibility and resistance to changes necessary for innovation support.

To effectively respond to these generic challenges, the organization must have Intrapreneurship Management that will align the perception of values and risks between management and Intrapreneur to attract, inspire, support, and reward

Intrapreneurs to develop and implement successful innovations as well as provide suitable tolerance for inevitable failures.

2.7 Major Changes in the Business and Natural Environment and Special Challenges to Intrapreneurship in the 21st Century

The integrated system analysis approach will next study major changes in not only the business but also the natural environments that impact the needs/wants and values of Innovation Team and Internal Supporter in an organization as well as the Adopter and, thus, pose special challenges to Intrapreneurship Development and Management in the 21st century.

2.7.1 Socioeconomic Developments

The Industrial Revolution from the mid-18th century to the mid-20th century resulted in the dominance of manufacturing industries in the economy. During this period, organizations focused on the physical productivity of employees as part of the manufacturing process. This led to the widespread acceptance of scientific management principles and practices by efficiency experts like Frederick Taylor [8], who viewed employees as an extension of production machinery. This approach used job specialization and time-motion studies to improve employee physical efficiency, and financial rewards were the primary motivational tools for employee productivity. The hierarchical structure and command-and-control management style had been the prevailing organizational culture, which had increased productivity, leading to enormous economic growth in industrialized countries. For example, real per capita GDP adjusted for inflation and purchasing power parities between 1750 and 1950 had increased more than 60-fold in the industrialized countries in spite of two major world wars and the Great Depression [9] and most people in these countries had literally been lifted from abject poverty to relative prosperity. Additionally, education levels expanded significantly among populations of these countries, with average literacy rate increasing from 10% to 90% during the 200-year period [2.10]. As a result, people in these countries, especially those born in the last 30 years who are beginning to enter the workforce, have now an unprecedented availability of physical, financial, and intellectual resources.

Starting in the mid-20th century, industrialized countries entered into the information age and gave rise to the service economy [11]. Employees in most organizations gradually transformed into knowledge workers, and organizations began focusing more on intellectual than physical productivity. The availability of financial and intellectual resources also elevated the security-oriented physical and psychological needs/wants to those for autonomy and excitement, fun, and meaning at work for workers entering into 21st-century organizations [12]. These paradigm shifts in the needs/wants and values of workers will result in fundamental changes in the development of Innovation Teams for organizations in the 21st century.

However, these shifts in needs/wants and values have also significantly changed the demands and expectations of the Final Adopters of innovations. They have increased both individualism and paradoxically faddish trends for many Final Adopters as they try to seek statements of their own individuality, but then follow fads *en masse* when they lack sufficient psychological maturity or intellectual resources to make truly individualistic self-statements. Moreover, Adopters now expect organizations and innovation to be socially responsible, environmentally conscious, and to provide personalized products (goods or services). They have also become more vocal in using social media and other online platforms to express their opinions and influence innovation developments. As a result, in the 21st century, the end-use markets for innovative products (goods or services) will be highly fragmented and fast-evolving, which will cause rapid rises as well as fast failures of many niche innovations, in the form of *mass customization* [13].

Finally, continuing from the 20th century, there have been increasing socioeconomic inequities in the world population, resulting in widespread regional conflicts and global unrest. These enormous socioeconomic and political problems have spawned many innovative but not all successful solutions, like communism and the United Nations. They will continue to pose important and urgent needs as well as challenges to innovation.

2.7.2 Technology Advances

Another major paradigm shift in the 21st century is the exponential growth of technologies, especially in information, communication, and biological sectors, which has resulted in the maturation of mobile communication, social media, robotics, artificial intelligence (AI), man–machine interactions, among others. These advances have already fundamentally transformed the way people communicate, work, and consume, and have also made major impacts on innovation management. Information technology has been used to implement crowdsourcing for creative idea generation [14], as well as to provide detailed assessments of employee personality and performance, allowing for targeted motivation, training, and career paths [15]. Similarly, information technology has also been used for crowdfunding [16]. Social media has been highly effective in both involving Adopters in idea generation [17] and enabling mass customization for individual need/want fulfillment [18]. Future technology developments will undoubtedly affect human cognition, perception, and decision-making, as in the examples of fake news and deepfake [19], which, in turn, will affect how various innovation participants can be marketed for the development and adoption of new products (goods or services).

Additionally, technological advances have enabled the creation of business model innovations, such as the sharing economy. Companies like Uber and Airbnb have disrupted traditional industries by offering convenient, affordable alternatives. However, these new models have also raised questions about regulation and labor rights.

On the other hand, with increasing economic well-being provided by innovations, human aspirations for independence and self-determination have expanded. Recent surveys [20, 21] of millennials, Gen-Xs, and Gen-Ys in advanced economies like those in North America and Western Europe as well as industrializing countries like China, show that the younger generations aspire for more autonomy and individualism, motivating them to be creative and innovative and increasing the popularity of innovation.

Moreover, the internet and social media have revolutionized marketing, making it easier and more cost-effective for businesses to reach a wider audience. Social media platforms like Facebook, Twitter, and Instagram provide businesses with a direct line of communication with their customers and allow them to build brand awareness, promote their products (goods or services), and engage with their audience.

Another major technological advancement that has impacted the business environment is AI. AI has the potential to transform how businesses operate by automating routine tasks, improving decision-making, and enabling new products (goods or services). For example, AI-powered chatbots are now used by businesses to automate customer service, while predictive analytics tools are used to analyze large amounts of data to make better business decisions.

2.7.3 Natural Environment Changes

The serious deterioration of the natural environment, caused by air and water pollution, land contamination, and other man-made disasters, has become a growing concern for individuals, organizations, and societies worldwide. Much of this has been the consequences of past innovations, such as undesirable outputs and side effects of industrial productions, agricultural and fishing practices, weapon developments, and medical treatments. In the highly connected world of the 21st century, such changes can quickly affect all people, as evidenced by the global impact of the recent COVID-19 epidemic. With growing awareness and concerns, there will be increasing regulations and incentives for different innovation developments, which will fundamentally affect the values and risks in the management of the innovation process.

Specifically, organizations will be increasingly required to adopt more sustainable practices, such as reducing carbon emissions, promoting renewable energy, and supporting ethical supply chains. Moreover, consumers as the Final Adopters of innovations have become increasingly conscious of the impact of their purchases and have shown a preference for sustainable products and companies. A 2020 study by Accenture found that 60% of consumers surveyed were willing to pay more for sustainable products, and 83% believed it was important for companies to design products that were meant to be reused or recycled [22].

2.8 Key Elements of Intrapreneurship Management

Based on the special interactions between Innovation Team and Internal Supporter, and their joint interactions with Adopter, together with the integrated system analysis of the Intrapreneurship innovation process, we can identify three key elements of effective Intrapreneurship Management, i.e., the effective development and implementation of innovative products (goods or services) in an organization.

The first key element is Organization Readiness for innovation, which assesses whether the organization has an innovative and productive culture, as well as whether it has a well-developed Internal Support system and an effective procedure for building an Innovation Team.

Once the Organization Readiness is assured, the other two interactive key elements are the Market and Technology Readiness of an innovation project and the resulting product. When all these three key elements reach high readiness levels, then the organization will have a high probability of developing and implementing significant impact innovative products. The systematic development of effective procedures for these readiness assessments and the decision tree analysis for either continuing to invest in improving the readiness, bringing the innovative product to market, or reallocating the resources to other projects will be the main objective of this book.

Summary

In this chapter, we have presented the innovation process as an interaction system of common key elements for implementing the products (goods or services) developed from a new or adapted creative idea to achieve impact. Major participants of the innovation process are Innovation Team, Supporter (including external supplier and regulator), and Adopter (including distributor and end-user). We have highlighted the special case for Intrapreneurship, i.e., Internal Innovation for an organization, with the Internal Supporter playing a pivotal role. We have then presented a common investment perspective for all participants and the importance for them to align their respective perceived expected values and risks for investing their time, energy, money, and other resources in the joint implementation of products (goods or services) developed from a creative idea. We have illustrated specific applications of the common investment perspective to individual participants. We have presented the generic challenges of Intrapreneurship due to misalignment of the investment perspectives between Innovation Team and Internal Supporter in an organization. We have next presented the special challenges to Intrapreneurship in the 21st century due to the impacts of major changes in the business and natural environment on the needs/wants and values of the innovation participants. Finally, we have identified three key elements for effective Intrapreneurship Management: Organization, Market, and Technology Readiness. The study of how to effectively assess and improve these three key elements is the main objective of the book.

Glossary

Adopter: Innovation participant who adopts products (goods or services) developed from a creative idea and includes Internal Adopter for Intrapreneurship as well as Distributor and End-User or Final Adopter.

Common investment perspective: A common perspective for balancing the perceived expected values and risks in investing time, energy, money, and other resources for implementing products (goods or services) developed from a creative idea to achieve significant impact; an alignment of the perspectives of all innovation participants is critical to the success of innovation implementation.

Innovation participants: Innovation Team, Supporter, and Adopter; for Intrapreneurship, they are Innovation Team, Internal Supporter, and Adopter.

Innovation process: A dynamic process that involves the implementation of products (goods or services) developed from a creative idea through an interactive and iterative system of common key elements of innovation motivation, needs identification, idea generation or adaptation, Innovation Team formation, proof of concept development, support seeking and provision, product refinement, organization expansion, and impact assessment, and re-innovation.

Innovation Team: A team that is formed and may gradually evolve for the development and implementation of a creative idea and resulting products (goods or services), and always has a champion dedicated to the success of innovation process; specifically, **Entrepreneur** represents an Innovation Team that forms its own independent organization with external support, while an **Intrapreneur** represents an Innovation Team inside an existing organization and is interchangeable with **Internal Innovator**.

Integrated system analysis approach: An approach to study the innovation process as an integrated and iterative system of the interactions among the innovation participants in implementing products (goods or services) developed from a creative idea to achieve significant impact.

Internal Innovation process: In this special case, the innovative organizational culture provides the innovation motivation as the starting point for Internal Innovation.

Key elements of Intrapreneurship Management: Organization, Market, and Technology Readiness.

Supporter: Innovation participant who provides mainly financial support to the innovation process, but may also provide sponsorship, mentoring, and other nonfinancial support, and also includes external supplier and regulator.

Discussions

- Critique the adequacy of the Internal Innovation process model and provide possible modifications and expansions.
- Discuss the acceptability and applicability of the investment decision perspective of the innovation participants in innovation development and implementation.
- Provide additional applications of the investment decision perspective for various innovation participants, such as the regulator and the distributor.
- Provide additional insights on the generic challenges of Internal Innovation.
- Review the major changes in the 21st century and their impacts on innovation needs/wants and challenges to innovation management.
- Review the three key elements of Intrapreneurship Management.
- Discuss the integrated system analysis approach for studying the interactions among major innovation participants and the organizational culture and potential assessment questions and checklists for measuring Organization, Technology, and Market Readiness.

References

1. Markowitz, H.M. (1952). Portfolio selection. *The Journal of Finance* 7: 77–91.
2. Elton, E.J., Gruber, M., and Brown, S. (2014). *Modern Portfolio Theory and Investment Analysis*, 9e. Wiley.
3. Elkington, J. (1998). *Cannibals with Forks: The Triple Bottom Line of 21st Century Business*. Capstone Publishing.
4. Manyika, J., Chui, M., and Miremadi, M. (2018). *The State of Innovation: Why Government Isn't Enough*. McKinsey Global Institute.
5. Christensen, C.M. (1997). *The Innovator's Dilemma: When New Technologies Cause Great Firms to Fail*. Harvard Business Press.
6. Bartz, W. (2019). *Global Innovation Index 2019: Creating Healthy Lives—The Future of Medical Innovation*. World Intellectual Property Organization.
7. Stam, E. (2018). Innovation management in the public sector: a systematic review and future research agenda. *Public Management Review* 20 (3): 441–463.
8. Taylor, F. (1911). *The Principles of Scientific Management*. Harper & brother.
9. Bairoch, P. (1995). *Economics and World History: Myths and Paradoxes*, 2e. University of Chicago Press.
10. Allen, R.C. (2003). Progress and poverty in early modern Europe. *The Economic History Review* 56 (3): 403–443.
11. Shelp, R. (1982). *Beyond Industrialization: Ascendancy of the Global Service Economy*. Prager Publisher.
12. Deal, J.J. and Levenson, A. (2016). *What Millennials Want From Work*. McGraw-Hill.
13. Kaplan, A.M. and Haenlein, M. (2006). Toward a parsimonious definition of traditional and electronic mass customization. *J. Product Innovation Management* 23: 168–182.
14. Brabham, D.C. (2018). *Using Crowdsourcing in Government*. IBM Center for the Business of Government.
15. Schweyer, A. (2018). *Predictive Analytics and Artificial Intelligence*. People Management, Incentive Research Foundation.
16. Falak, D., Shanawaz, S., Pranav, J. et al. (2022). Crowd-funding using blockchain technology. *International Journal of Research Publication and Reviews* 3 (11): 2214–2216.
17. Martinez, M. (2018). *Xiaomi Creates Client Loyalty Through Cocreation in Their Customer Journey*. Idea4all Innovation.
18. Fatur, P., Novak, B., and Dolinsek, S. (2017). Mass customization in footwear industry: a case study. *Proceedings of the 8th International Conference of the Management*, Australia (17–19 November 2017), pp. 1383–1389.

19 Galston, W.A. (2020). *Is Seeing Still Believing? The Deepfake Challenge to Truth in Politics*. Brookings Institution.

20 Deloitte (2018). Millennial survey. https://www2.deloitte.com/content/dam/Deloitte/global/Documents/About-Deloitte/gx-dttl-2018-millennial-survey-report.pdf.

21 IBM Institute for Business Value (2016). Redefining boundaries: insights from the global C-suite study. https://www.ibm.com/downloads/cas/4EXLNJZW.

22 Accenture (2020). Consumers say sustainability is a factor in purchasing decisions. https://www.accenture.com/_acnmedia/PDF-113/Accenture-Consumer-Study-Sustainability.pdf.

3

Understanding Human Needs and Wants as Driving Forces of Intrapreneurship

Intrapreneurship involves the joint investments of time, money, effort, and resources by the Intrapreneur, Internal Supporter, and Adopter in attaining their individually perceived expected values with acceptable perceived expected risk through the development and implementation of innovation in an organization. These perceived expected values and risks originate from individual human needs and wants but can be extended to represent the collective or aggregate needs and wants of a group representing Innovation Team, Internal Support Board, or Adoption Committee. In this chapter, we will present an overview of human needs and wants and discuss their characteristics and applications to Intrapreneurship. We will also present a new perspective on the interactions among these needs and wants and extend them from individuals to organizations and societies in general to develop an integrated value framework for innovations.

3.1 Overview of Human Needs and Wants

Human needs and wants are highly complex, shaped by a combination of biological, social, cultural, and experiential factors. The more urgent needs go beyond physical survival and subsistence and encompass physical and psychological safety, security, stability, and a sense of well-being. However, individuals have less pressing but highly diverse wants or desires for physical and psychological stimulations, pleasure, status, growth, and fulfillment in an interactive biological, social, and cultural environment.

Intrapreneurship Management: Concepts, Methods, and Software for Managing Technological Innovation in Organizations, First Edition. Rainer Hasenauer and Oliver Yu.

Published 2024 by John Wiley & Sons, Inc.

The availability of resources can also affect an individual's awareness and ability to satisfy their different needs and wants. Specifically,

- Economic resources, such as money and job opportunities, can provide access to physical and intellectual resources, allowing individuals to satisfy basic physical needs for food, water, shelter, and health care, as well as wants for comfort, convenience, or leisure activities.
- Intellectual resources, such as specialized training or general education in critical thinking, are essential for uncovering and satisfying higher level needs, such as personal growth and search for life meaning. These resources also enable individuals to acquire new skills, knowledge, and perspectives to help satisfy their wants and desires for pursuing their passions and interests and for enjoying aesthetic stimulation or cultural experiences.
- Belief-based resources, such as religion, ideology, and spirituality, can affect an individual's sense of purpose and meaning in life. Access to these resources can also provide individuals with a sense of community, belonging, and shared values, which can be essential for satisfying their social and emotional needs. Belief-based resources can also facilitate the satisfaction of wants, such as the desire for spiritual growth or self-discovery.

Based on leading studies and surveys on human needs, wants, desires, and motivations [1–14], we have compiled a list of significant primary human needs and wants and classified them into broad categories based on the apparent security or stimulation-seeking nature and physical or psychological characteristics, as shown in Table 3.1.

Table 3.1 Classification of Significant Primary Human Needs and Wants, Relations to Resource Availability, and Applicability to Value Creation for Intrapreneur and Internal Supporter in an Organization

Significant Human Needs and Wants	Needs for Security and Stability		Wants for Stimulation and Growth		Relation to Resource Availability	Application to Internal Innovation
	Physical	Psychological	Physical	Psychological		
Physical Survival: Air, Water, Food, Rest	x	x			E	
Physical Safety and Protection	x	x			E	

Table 3.1 (Continued)

Significant Human Needs and Wants	Needs for Security and Stability		Wants for Stimulation and Growth		Relation to Resource Availability	Application to Internal Innovation
	Physical	Psychological	Physical	Psychological		
Health Care and Clean Environment	x	x			E, INT	
Mental Well-Being and Emotional Stability	x	x		x	E, INT	
Comfort, Conveniences, and Luxury			x	x	E	
Sensory Stimulation and Enjoyment	x	x	x	x	E	
Recreation and Leisure Activities	x	x	x	x	E	
Love and Affection	x	x	x	x	E	
Family Formation and Procreation	x	x	x	x	E	
Managing Unknowns with Either Beliefs or Explorations		x		x	INT, BB	I
Learning and Intellectual Stimulation		x		x	INT	I
Creativity and Self-Expression				x	INT	I
Challenge and Competition		x		x	INT	I
Friendship and Social Interaction		x		x	INT, BB	I
Group Affinity, Acceptance, and Belonging		x		x	INT, BB	I, S
Stability and Order		x		x	INT, BB	I, S

(*Continued*)

Table 3.1 (Continued)

Significant Human Needs and Wants	Needs for Security and Stability		Wants for Stimulation and Growth		Relation to Resource Availability	Application to Internal Innovation
	Physical	Psychological	Physical	Psychological		
Collaboration and Contribution		x		x	INT, BB	I, S
Empathy, Fairness, Equity, and Justice		x		x	INT, BB	I, S
Attention, Popularity, and Fame		x		x		I
Sense of Achievement and Satisfaction		x		x		I, S
Recognition, Respect, and Self-Esteem (Avoidance of Failure and Shame)		x		x		I, S
Individual Autonomy and Freedom		x		x	INT	I
Financial Rewards, Security, and Independence		x		x	E	I
Status and Privileges		x		x		I, S
Power and Control		x		x	INT	I, S
Personal Development and Growth				x	INT, BB	I

BB, belief-based; E, Economic; INT, intellectual; I, Intrapreneur; S, supporter.

In this table, we have employed subjective judgment to identify the primary relationship between these needs and wants to the availability of economic, intellectual, and belief-based resources. All these needs and wants are directly applicable to creating values for the Adopter through their fulfillment. We further used studies on employee motivations to identify the needs and wants that are particularly applicable to internal innovation participants in an organization. Finally, we again used subjective judgment to identify the needs and wants applicable to creating values for the Internal Supporter, based on its frequent dual role as the management of an organization.

It is important to note that due to the complexity of human needs and wants and the interconnectedness between their physical and psychological aspects, this table is not exhaustive or definitive but a representative collection and a simplified, subjective, but reasonable, categorization based on the authors' collective experience. Readers are encouraged to use their own experience and judgment to modify or supplement this table.

3.2 General Observations of Human Needs and Wants

From Table 3.1, we can make few general observations of human needs and wants.

- *Human needs and wants are diverse and often appear to conflict with each other and their emergence are not necessarily hierarchical but can be discontinuous.*

As individuals gain economic resources and access to additional intellectual and belief-based resources beyond those required for satisfying basic physical needs, their needs and wants start to diversify. These diverse needs and wants often appear to be in conflict with one another, such as the need for order and stability versus the desire for individual autonomy and freedom, or the need for collaboration versus the want for competition. Moreover, these needs and wants do not necessarily emerge in a hierarchical manner, but they can be discontinuous, as seen in cases where deeply religious or ideologically indoctrinated individuals may sacrifice their basic needs for physical survival to directly satisfy their wants for life meaning or spiritual fulfillment.

However, such apparent conflicts and discontinuities may stem from the inherent tension between human needs for security and stability and wants for stimulation and growth, both of which are fundamental to individual survival and species continuation. People generally strive for a balance or make trade-offs among these needs and wants based on their personalities and other characteristics. For example, a conservative person may need more order and stability, whereas a driven person may want more individual autonomy. A sociable individual may desire more collaboration, whereas an aggressive person may want more challenge and competition. In addition, a person may sacrifice some degree of the want for self-expression in exchange for fulfilling the need for security in the form of acceptance by a group, or vice versa.

Another example of this balance can be seen in the reception of new products by Adopter. Research [15] has shown that a product is generally best received by the customer if it is new but not too radically different from a similar existing product, reflecting the customer's balance between the want for the stimulation of experiencing a new product and the need for the security from avoiding the risk of failure and the resulting trouble in using a highly unfamiliar, new product.

However, many individuals believe that they can satisfy both their needs for security and wants for freedom through the possession of wealth, power, healthy immortality, or inner peace and well-being derived from spiritual wisdom, religious fervor, or ideological zeal. Such beliefs have been driving forces for popular pursuits of wealth, power, medical youth, as well as many religious sects and ideological doctrines.

- *There are also complementarity, substitution, and transmutation among needs and wants.*

In addition to conflicts, human needs and wants also have complementarity, substitution, or transmutation characteristics. For example, the need for collaboration for psychological security can complement the need for psychological stimulation and growth through collaboration. However, for many individuals, the desire for sensory pleasures to satisfy the need of physical stimulation can often be substituted (or sublimate into platonic pleasures) by the desire for love and affection to satisfy the need for psychological security. Moreover, while physical wants for stimulation and growth typically follow the satisfaction of physical needs for security and stability, few individuals may feel more confident and secure when they are free and uncontrolled by others, whereas others may feel freedom from worries when they are under total protection and control by powerful authorities.

The transmutation between the need for security and the want for stimulation and freedom is also noteworthy. For instance, alcohol, tobacco, and other drug use, as well as gambling or social media activities, may initially fulfill the want for stimulation but can later transform through addiction into fulfilling the need for psychological dependency as a form of security. This highlights the unethical motivation for some marketers to make a product addictive. Another example is acceptance by a group, which can initially fulfill the need for both physical and psychological security, but can also lead to individual behaviors that fulfill the wants for psychological stimulation from life meaning or spiritual freedom.

- *Economic resources play a major role in meeting basic physical needs, but intellectual and belief-based resources are increasingly important to a wide range of diverse needs and wants.*

Table 3.1 indicates that economic resources play a major role in meeting basic physical needs, whereas historical belief-based resources of tradition and religion are important in fulfilling psychological needs for the search for answers and life meaning. However, as economic resources increase, intellectual resources tend to expand as well, resulting in the emergence of many intellectually based wants, such as life meaning from contribution to the well-being of the world population or future generations and devotion to an ideology. As a result, in addition to the value of economic prosperity in satisfying the needs for financial security and the wants for luxury, sensory, and other materialistic pleasures, values in satisfying the needs and wants for environmental quality and social equity have become increasingly important for people in the 21st century.

Nevertheless, it is still essential to be aware of the dominance of economic resources in human needs' evolution through history and in many parts of the world today. Until the industrial revolution in the 1750s, and even for 90% of the world population today, economic resources have been scarce. As a result, the needs for sufficient supply of life essentials and adequate living standards remain very strong for most people. These needs become even more urgent during times of economic hardship due to economic recessions and human-made or natural disasters. Furthermore, for people who do not possess sufficient economic resources, needs for monetary rewards and financial security will continue to be of predominant importance.

However, intellectual and belief-based resources have complex relationships. While intellectual resources provide people with analytical, critical, and creative thinking capabilities, belief-based resources support a person's ability to accept and adhere to preset beliefs. Thus, people with more intellectual resources may have more degrees of freedom to pursue new ideas and creative thinking to satisfy their psychological wants for stimulation and growth. In contrast, those with rich belief-based resources, such as strong religious or ideological beliefs, may have more degrees of freedom to participate in collaborative efforts within like-minded communities to satisfy their differently perceived psychological wants for stimulation and growth. It is worth noting, however, that intellectual resources are largely developed through education, which can also be a major approach to belief forming. Therefore, biased perceptions and misguided thinking processes can sometimes be built into intellectual resources, making them similar to belief-based resources.

Finally, as both intellectual and belief-based resources are affected by the type and degree of belief-inducing activities, including biased education, indoctrination, propaganda, and advertisement, which have strong national and cultural variability, the emergence of human needs and wants related to these resources, thus, also have large national and cultural differences. Combining this observation with the previously discussed individual balance between the needs for security and wants for stimulation, there are also important national and cultural differences to be taken into consideration in understanding specific individual's needs and wants.

3.3 Major Factors Influencing the Intensity of Needs and Wants

The trade-off between conflicting needs and wants is dependent on the relative intensities of various needs and wants, which are in turn influenced by several major interrelated factors:

- *Resource availability and personality traits.*

In addition to the influences of economic, intellectual, and belief-based resources on the intensity of many physical and psychological needs and wants beyond the basic needs for physical survival, research has shown that personality traits also have significant influences on the intensity of needs and wants. In particular, extensive studies [16–20] based on the Big-Five personality traits have indicated the following influences on needs and wants:

Openness to experience: Individuals who are open to new experiences tend to have a higher intensity of needs and wants related to exploration and creativity.

Conscientiousness: Conscientious individuals tend to be organized and goal-oriented, which can influence the intensity of needs and wants related to achievements.

Extraversion: Extraverted individuals tend to have a higher need for social interaction and may seek out experiences that provide social stimulation. This can influence the intensity of needs and wants related to friendship and social interactions with others.

Agreeableness: Agreeable individuals tend to value cooperation and harmony, which can influence the intensity of needs and wants related to social interaction and collaboration.

Neuroticism: Neurotic individuals tend to experience negative emotions more frequently and intensely, which can influence the intensity of needs and wants related to emotional well-being.

On the contrary, the opposites of these personality traits will similarly influence the intensity of the contrasting needs and wants.

- *Social environment and indoctrination*

Another set of interrelated factors affecting the intensity of needs and wants are characteristics of the social environment and various forms of indoctrination, which include all types of education, training, and propaganda carried out through different media. The characteristics of the social environment include culture and tradition that directly shape the intensity of needs and wants of individuals. Specifically, in a social environment, individuals learn the norms, values, and behaviors of their culture and tradition. For example, in some cultures, there is a strong emphasis on material possessions, which can lead to a higher intensity of wants for luxury goods and services, whereas in other cultures, there may be a strong emphasis on self-restraint and frugality, which can lead to a lower intensity of wants for material possessions.

Individuals are also formally or informally indoctrinated in a social environment to acquire attitudes and beliefs about what they should need and want, which can influence the intensity of their actual needs and wants. For example, for most of the regions in the world, large number of individuals are indoctrinated through formal education or informal propaganda to value certain needs and wants and disdain others. Even more clearly about the impact of these factors is the well-documented success of commercial advertising [21] in arousing the intensity of a need or want that people may not have been even previously aware of.

3.4 Special Applications to Intrapreneurship

All the needs and wants are applicable to the Adopter in the internal innovation process of an organization as a representative human customer. However, there are special applications of these needs and wants to the Intrapreneur and Internal Supporter in an organization.

- *To the Intrapreneur*

The tension between the stimulation and insecurity from a new idea applies strongly to the Intrapreneur. It also provides a basis for differentiation between Entrepreneur and Intrapreneur. Both Entrepreneur and Intrapreneur have desires for intellectual curiosity and excitement, creative pursuits and fantasies, individual autonomy and freedom, as well as sense of achievement and contribution; however, the intensities of these desires are different. Specifically, Entrepreneur is strongly driven by these desires to the degree that innovation is like a “life calling” [22], whereas these desires are less intense for Intrapreneur as they are moderated by the needs for Group Affiliation and Belonging, Stability and Order, and Financial Security. As a result, the balance of tension between stimulation and insecurity for Intrapreneur is toward moderate value aspiration and moderate risk tolerance, as discussed in the “Investment Perspectives” in Chapter 2.

The observation that the needs for security and stimulation can be complementary and substitutable is also of special importance to Intrapreneur. For example, the desire for collaboration in one area can complement the want for competition in another area and result in “coopetition,” which is particularly important for innovation management in case of joint venture for technology development. A case example is the joint venture of Sony Corporation and Samsung Electronics for the development and manufacture of flat LCD TV panels, known as S-LCD. Because of the S-LCD cooperation on the one hand, the whole LCD industry prevailed over the PDP technology in flat TV panels. This led to a quick advancement of PDP and the LCD technology because of the ensuing competition within the industry on the other hand. S-LCD was further responsible for the drop in the price of flat-screen TVs, due to the competitor’s response and the economies of scale [23].

- *To the internal supporter*

The tension between stimulation and insecurity about a new idea also applies to the Internal Supporter as representative of the management of an organization. Although the External Supporter is largely driven by the desire for Financial Rewards, the Internal Supporter has a much more complex balance among his or her desires. Specifically, the Internal Supporter needs to balance between the wants for Intellectual Excitement, Achievement, and Financial Rewards from a promising internal innovation and the competing desires for Financial Rewards from a successful existing business together with the needs for Stability and Order within the organization as well as Avoidance of Failure and Shame. The inability to maintain an effective balance between these desires for Internal Supporter has been the key difficulty for developing Intrapreneur in an established organization, and its effective resolution is the primary objective of this book.

However, combining with attractive financial incentive and effective personality selection, an innovative culture and powerful training programs can strongly intensify the needs and wants for innovation for both Intrapreneur and Internal Supporter in an organization. More importantly, Internal Supporter must realize that in a world of fierce competition, not only business growth but even organizational survival requires continuous innovation. It is literally Innovate or Perish.

3.5 Graphical Representations of Human Needs and Wants: The Maslow Hierarchical Model and a New Perspective

Graphical representations of human needs and wants can often be useful in communicating ideas. A widely recognized example is Abraham Maslow's Hierarchy of Needs, as shown in Figure 3.1.

Abraham Maslow [1] has made groundbreaking advances in showing that the fulfillment of many needs, such as self-esteem and self-actualization, requires preconditions of satisfying the physiological, safety, and love/belonging needs. However, in practice, behaviors such as seeking stimulation or growth, which are intermediate between basic survival needs and the realization of high ideals, are expressions of psychological needs but do not necessarily precede or succeed one another. Hence, the needs "hierarchy" has middle layers that are far from fixed. In particular, even self-esteem needs do not necessarily follow love/belonging needs. In fact, as discussed earlier, these needs can occur in parallel and even be complementary or substitutable to one another. Nevertheless, the pyramidal shape of the model is often misconstrued as indicating that human needs are one-directionally hierarchical. To avoid this misimpression, a more practical, new perspective can be developed by separating the needs and wants into physical and psychological dimensions as well as security and stimulation dimensions. Based on these separations, we have developed the following new graphical representations of the new perspective on human needs and wants for individuals as well as organizations and societies as a whole.

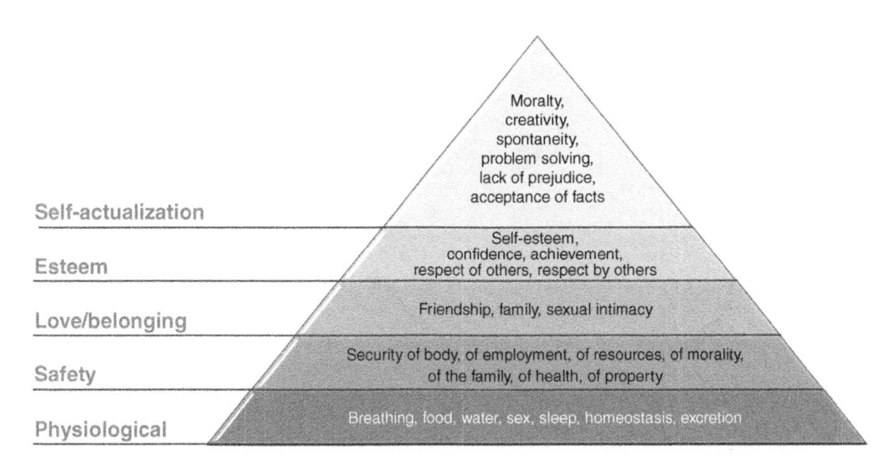

Figure 3.1 Maslow's Hierarchy of Human Needs

3.5.1 Graphical Representation of a New Perspective on Individual Human Needs and Wants

Based on the discussions in the earlier sections of this chapter, we propose the graphical representation of a new perspective on individual human needs and wants based on the physical/psychological and security/stimulation dimensions, as shown in Figure 3.2.

In this new perspective, individual needs and wants can be categorized into four combination clusters based on the dominance of needs or wants. The needs and wants in each cluster can be arranged into a loose hierarchy in accordance with increasing availability of physical, economic, intellectual, and belief-based resources, such that as resources increase, higher level needs and wants emerge and expand.

For the *physical-security* cluster, the needs and wants are largely focused inward on the physical security of an individual. They start at the lowest level with physical survival needs and expand to needs for physical safety, health, and mental well-being. With increasing availability of resources, they expand to include wants for comfort, convenience, and luxury.

For the *psychological-security* cluster, the needs and wants are more diverse and complex and largely focused inward on the psychological security of an individual

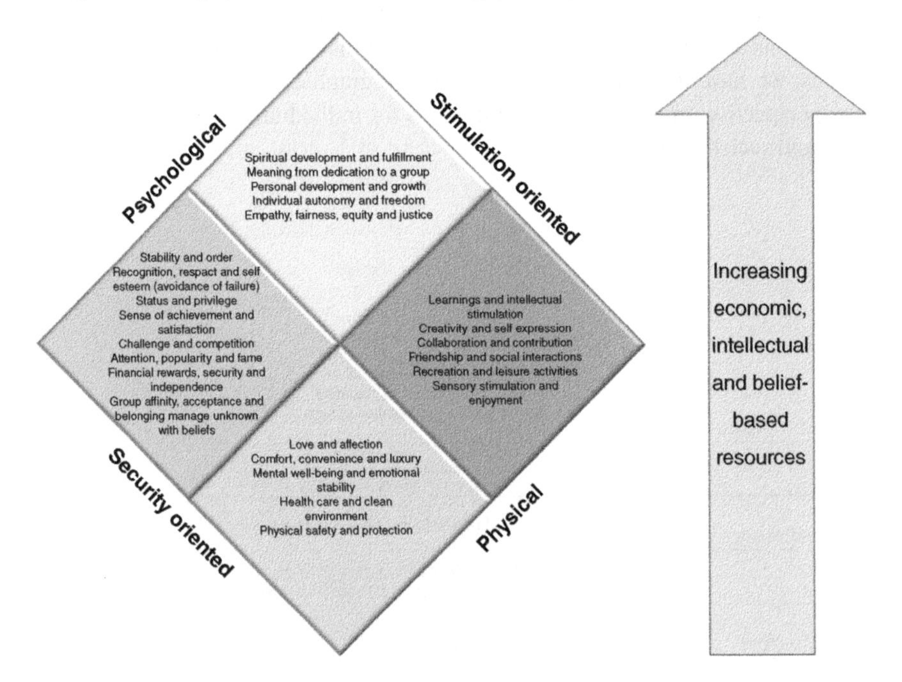

Figure 3.2 Graphical Representation of a New Perspective on Human Needs and Wants

with respect to the physical and social environment. They start with the need to manage the unknown, which often manifests as the need for religious beliefs, including superstitions, and expands to the needs for group affinity, acceptance, and belonging. They further expand to the needs for stability and order, self-esteem, achievement, and power and control.

For the *physical-stimulation* cluster, the needs and wants are again diverse and complex and more outwardly focused on either the enjoyment from interacting with others or with own imagination, creativity, and enlightenment. They start with sensory stimulation and pleasures, such as tasty food and enjoyable sex, and expand to friendship and social interactions. With increasing resources, they further expand to the wants for recreation and leisure, and for intellectual curiosity to probe and understand the unknowns and creative pursuits for new enjoyment and enlightenment in the psychological-stimulation cluster above.

Last, for the *Psychological-Stimulation* cluster, the needs and wants are largely outwardly focused on empathy with others, meaning of life, and relation with the universe. They start with empathy, fairness, equity and justice, and individual autonomy and freedom and expand to personal development and growth, and to life meaning through contributions and spiritual development and growth with increasing intellectual and belief-based resources.

The diamond shape of the graphical representation also offers the following implications: The basic physical survival needs are narrowly focused and intense, whereas the subsequent needs and wants quickly expand and diversify with increasing economic resources. Yet, due to the requirement of superabundant intellectual and/or belief-based resources, the top level wants and desires for meaning and spiritual development and growth become more focused and less aspired.

This categorization of needs and wants is by no means absolute. Many needs and wants can cross over from one cluster to another. For example, the want for creative pursuits can be a part of the need for self-esteem, which can be the beginning of the need for personal development and growth. Similarly, the need for stability and order can overlap with the want for collaboration and contribution. Moreover, the loose hierarchies of the needs and wants among the clusters can also be permeable. For example, a person may transcend a basic need in the physical-security cluster, such as safety and health, to the wants in another cluster, such as recreation in the physical-stimulation cluster or self-esteem in the psychological-stimulation cluster.

Even though the new perspective of needs and wants does not produce a perfect orderly categorization, it can provide useful insights to individuals on the interactions among various needs and wants and how the intensities of these needs and wants can be influenced by resource availability, personality traits, social environment, indoctrination, and other factors.

3.5.2 Application of the New Perspective to Principal Participants of Intrapreneurship

The new perspective applies to the general needs and wants of all innovation participants. Specifically, Figure 3.2 is fully applicable to the Adopter as a representative human. We can also apply the new perspective to the categorization of needs and wants of the Intrapreneur and Internal Supporter for internal innovation in an organization, as indicated in the last column of Table 3.1 and shown in Figures 3.3 and 3.4, respectively.

Here, the Intrapreneur is more likely focused on a Secure and Pleasant Work Environment, Group Affinity, Financial Rewards, Attention, Challenge, Achievement, Recognition, Avoidance of Failure, Collaboration, Creativity, Intellectual Stimulation, Sense of Fairness and Equity, Personal Development, Meaning from Contribution to the Group, and potential Spiritual Fulfillment.

In contrast, as representative for the organization, Internal Supporter is more likely focused on Financial Rewards, Recognition, Avoidance of Failure, Stability and Order, Power and Control, Intellectual Stimulation, Empathy and Fairness, Personal Growth, and Contribution to the Group.

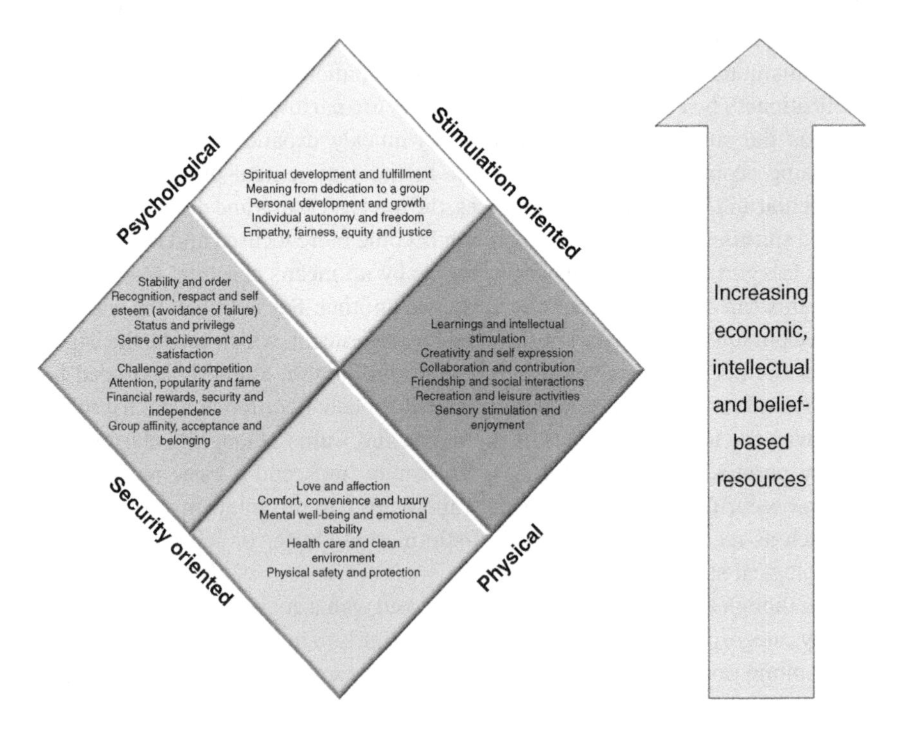

Figure 3.3 Graphical Representation of the New Perspective on Intrapreneur

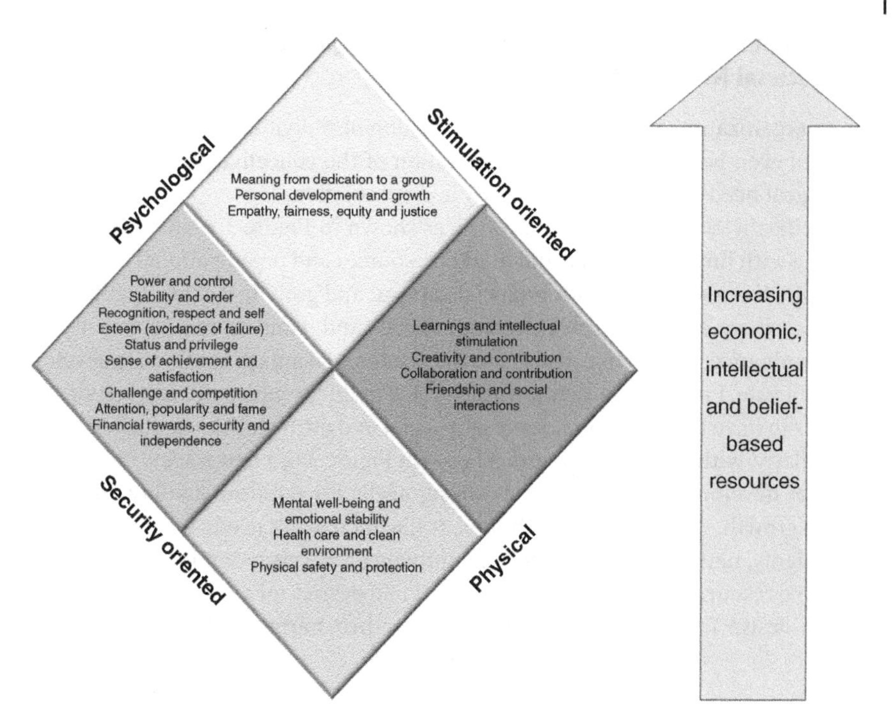

Figure 3.4 Graphical Representation of the New Perspective on the Internal Supporter

In summary, although the categorizations are relatively coarse and highly generalized, the new perspective does provide basic insights and a starting point on how to explore and understand the major needs and wants of the principal participants of the internal innovation process of an organization. Based on the exploration and understanding, the organization can develop incentives and guidance for effective Intrapreneurship development and management. However, it is important to note that exploration of the needs and wants through simple surveys or even focus group discussions based on this new perspective is often not sufficient because of the general superficiality and prevalent bias of the questionnaires and group discussions. A proven useful tool for determining these specific needs and wants of the innovation participants is embodied in the initial step of the Design Thinking approach to creative problem-solution. By this approach, to understand a person's needs and wants, one must practice empathy, that is, how to think and feel in the role of the person, observe his/her actual behaviors, and have meaningful discussions with the person to verify and refine the understanding of his/her needs and wants. Additional details on this approach will be provided in Chapter 5.

3.5.3 Extensions of the New Perspective to Organizational and Societal Needs and Wants

As an organization and a society are a collection of individuals, the new perspective can even be applied to the categorization of the collective physical and psychological needs and wants for security and stimulation.

Specifically, with the new perspective, as shown in Figure 3.5, an organization starting with limited financial and market resources and organizational capability generally has a strong need for financial survival and growth. As the organization becomes more established, it extends the needs and wants to market expansion and dominance; meanwhile, it also has the needs and wants to continue to develop and innovate. Finally, when achieving a level of maturity with sufficient resources, it starts to have the wants and desires for social and even global responsibility.

Similarly, with this perspective, as shown in Figure 3.6, a new society at an early stage of development generally has strong needs for political stability and economic growth. As the society stabilizes, it has increasing needs for independent sovereignty, national security, and local influence, and even dominance. Then, as available resources increase, it has the wants and desires for social equity followed by the desire for international dominance and/or harmony dependent on the

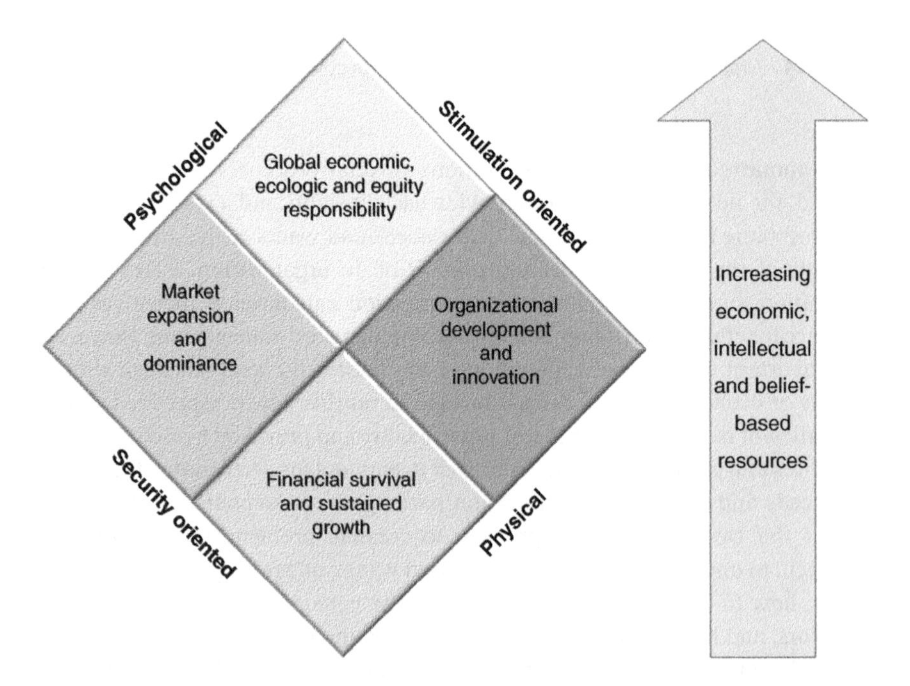

Figure 3.5 A Simplified Perspective of Organizational Needs and Wants

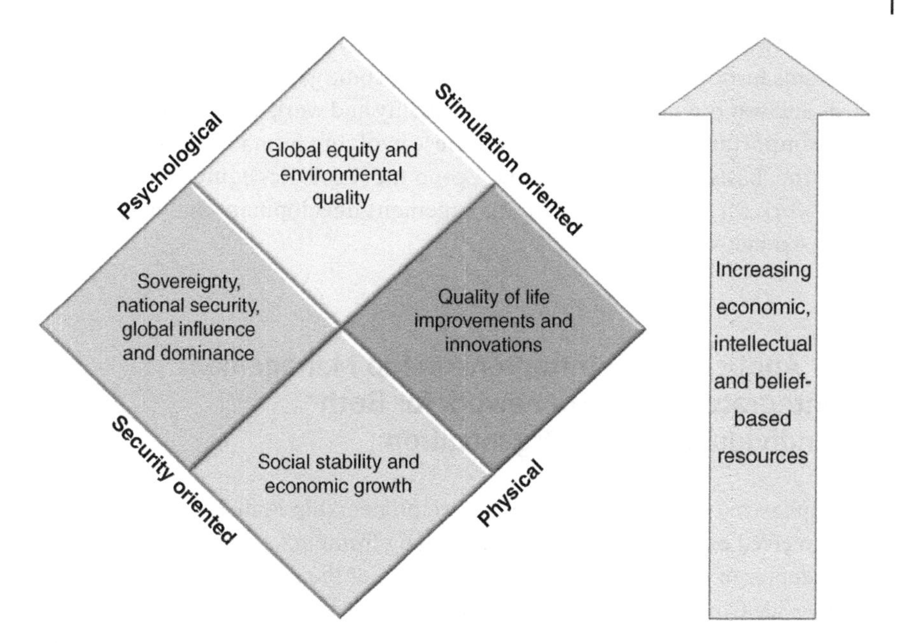

Figure 3.6 A Simplified Perspective of Societal Needs and Wants

perceived needs and wants of the representative society leaders. The evolution of the United States and the People's Republic of China since their respective foundings in 1776 and 1949 can be viewed as representative cases of this perspective.

These extensions can be useful in studying how an organization or a society would perceive the values and risks for the investment of its limited resources as supporters to the implementation of innovative products.

The following presents another real-life case example of this new perspective on expanding societal needs and wants with increasing economic resources: Between 1995 and 2010, one of the authors of this book was commissioned to assist in the development of 2 billion US dollars per year, in 1995, rolling 10-year national technology investment plan for a newly industrialized country in Asia. The starting point for the plan was the projected future needs and wants of that country. For many years prior to 1995, this country had been focusing almost exclusively on economic growth. However, when the project started in 1995, the per capita GDP had reached 12,300 US Dollars in 1995 or about 24,000 US Dollars in 2023 providing ample economic resources for most of the population in the country. More than 60 civic leaders including top government officials and prominent industrialists, scientists, technologists, educators, and artists were polled that year for the project. The consensus was that economic growth represented only about 50% of the country's needs and wants. The other 50% is split evenly between the needs and wants for social equity and quality of life. The needs and wants for social equity included reducing income gaps, whereas quality of life needs and wants included environmental quality and work–leisure balance.

It is important to point out that all these evolving human needs and wants form the basis for the integrated Economic–Ecologic–Equity (EEE) value framework for Intrapreneurship Management development and assessment discussed below.

3.6 Application to Intrapreneurship Management: An Integrated Value Framework for Both the Individual and the Organization

As emphasized throughout this book, Intrapreneurship Management is to *align* the perceived expected values and risks among Intrapreneur, Internal Supporter, and Adopter in investing in the implementation of the product developed from a creative idea to achieve significant impact. Understanding human needs and wants from the new perspective can be used to develop an integrated framework for the continuing balance among these individually perceived expected values and risks for innovation as the organization and the society evolve.

Discussions in Chapter 2 have detailed the great expansions in economic and intellectual resources over the last two centuries for developed nations, such as those in North America and Western Europe, as well as many newly industrialized countries, such as China, South Korea, and India. These increasing economic and intellectual resources have facilitated the global awareness of the serious deterioration of the natural environment, including air and water pollution and land contamination due to past innovations, and aroused strong human physical needs for a clean environment. Meanwhile, they have also expanded the human psychological wants for empathy, fairness, equity, and justice for fellow human beings. These rising needs and wants have begun to pervade the institutions and organizations of these countries. As a result, there have been increasing government regulations and incentives as well as public pressures and worker demands for organizations to reduce negative environmental impact, improve social equity, and promote corporate ethics. In response to these expanding needs and wants of both external customers and internal staff, the value of organizations has steadily shifted from profit maximization to a balance with social responsibilities, which include both environmental and societal contributions.

In 1998, British sociologist and environmentalist, John Elkington proposed the concept of the Triple Bottom Lines (TBL) [24] for long-term sustainability of an organization, which considers not only the traditional economic values but also the ecologic and equity values of the business development of the organization. In fact, since the start of the 21st century, the Environment, Social, and Governance (ESG) risk score has been increasingly used around the world to measure sustainability and societal impacts of an investment by a business organization. In response to these evolving market needs, organizations in the world through either active self-enlightenment or passive compliance to external pressure have gradually expanded from the historical economic-centric fixation to include long-term sustainability and equity-based business models and marketing strategies.

TBL has generally focused on the minimum requirements of these values. Dyllick and Hockerts [25] expanded the TBL concept to a strategy for balancing the Triple Top Lines (TTL) consisting of the business case for Profit, the natural case for Planet, and the societal case for People to achieve corporate sustainability, as shown in Figure 3.7. With rising needs and wants for environmental and societal responsibilities among individuals, organizations, and societies around the world, we have applied the new perspective on human needs and wants to further expand the work of Dyllick and Hockerts to develop an integrated economic-ecologic-equity (EEE) value framework for innovation assessment. This integrated assessment framework focuses on *maximizing* the combined EEE value, or the TTL of an innovative product, subject to the TBL requirements.

Specifically, both an innovative organization and innovation project need to continuously consider the EEE values in their respective development, evolution, and assessment. Such consideration is totally consistent with the UN Sustainable Development Guidelines (SDG) and the circular economy for sustainable growth,

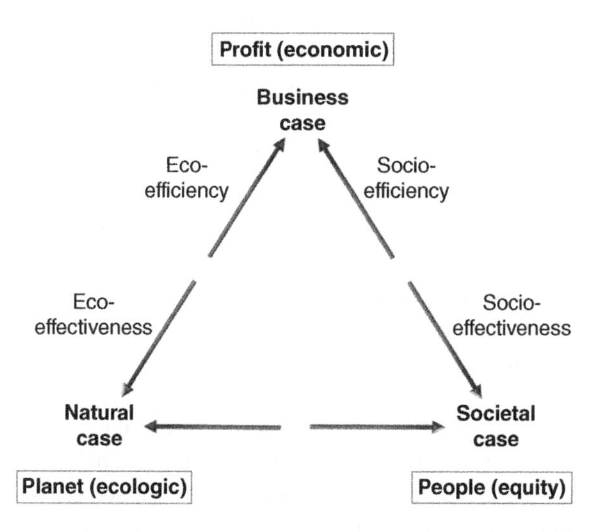

Figure 3.7 The Triple Top Line Business Strategy Model based on Dyllick and Hockerts

which involves the positive interactions of the EEE values for an organization and a society. The integrated EEE value framework is thus the basis for assessing the organization, market, and technology readiness of an innovation project developed by an organization in the IntrapreneurshipREADINESSnavigator (IRN©) software.

Summary

This chapter provides an integrative perspective on human needs and wants as consisting of interactive physical and psychological needs and wants for security and stimulation. These needs and wants evolve and expand with increasing economic, intellectual, and belief-based resources, but become more focused with superabundant intellectual or belief-based resources. These needs and wants can be used to estimate the perceived expected values and risks for the innovation participants in their decisions to invest their time, money, effort, and other resources to implement the product developed from a creative idea to achieve significant impact. Specifically, with rapidly evolving economic, technological, and environmental changes in the 21st century, it is critically important for Intrapreneurship Management in an organization to apply new perspectives on needs and wants and the integrated EEE value framework to effectively align and motivate innovation development and implementation investments among all participants inside and outside the organization. In addition, understanding the perception, thinking, and belief-forming processes can be used to develop creativity skills and personality transformation tools for innovation participants. These discussions can be further used to form the basis of customized questionnaires and checklists for the IntrapreneurshipREADINESSnavigator (IRN©) software for assessing the organization, technology, and market readiness of a specific innovation project of an organization.

Glossary

A new perspective on human needs and wants: This new perspective categorizes human needs and wants into combinations of physical–psychological and security–stimulation dimensions.

Hierarchy of human needs and wants: There is a loose hierarchy among human needs and wants, with the physical and psychological survival needs at the bottom, which expand with increasing economic, intellectual, and belief-based resources into diverse needs and wants for physical and psychological stimulation, including comfort, convenience, recreation, love, respect, and so on, and finally into more focused wants for meaning and fulfillment at the top.

Integrated EEE value framework: Based on continued global economic expansion since 1750 and the evolution of human needs and wants in the developed and newly industrialized world where physical survival needs have largely been satisfied, two new values – ecologic sustainability and societal equity, have emerged and gained increasing importance to form an integrated Economic–Ecologic–Equity (EEE) framework for assessing the values of individuals, organizations, and societies.

Interactions among needs and wants: Needs and wants can conflict with each other, complement or substitute each other, and transmute from one to another.

Need: Necessary for physical or psychological survival.

Triple Bottom Lines (TBL): The minimum requirements for Economic, Ecologic, and Equity values.

Triple Top Lines (TTL): The optimization of Economic, Ecologic, and Equity values.

Want or Desire: Beyond the needs and the two terms are interchangeable.

Discussions

- Study online, report, and discuss in class major theories of human needs, personality traits, perception, thinking, and decision-making processes.
- Critique the basic human needs for physical and psychological security and stimulation.
- Critique the respective new graphical models of human needs for individuals, organizations, and the society as a whole.
- List any missing major wants and desires of Intrapreneur and Internal Supporter and discuss these new ones with those in Table 3.1.
- Discuss your experience in motivating employees in an organization to be Intrapreneurs.
- Discuss your experience in motivating managers in an organization to be Internal Supporters.
- Discuss your experience in motivating customers to adopt innovative product or service.

References

1 Maslow, A.H. (1943). A theory of human motivation. *Psychological Review* 50 (4): 370–396.

2 Reiss, S. (2004). Multifaceted nature of intrinsic motivation: the theory of 16 basic desires. *Review of General Psychology* 8 (3): 179–193.

3 McClelland, D.C. (1961). *The Achieving Society*. Van Nostrand.

4 United Nations (2015). Transforming our world: the 2030 agenda for sustainable development. https://sustainabledevelopment.un.org/post2015/transformingourworld

5 Inglehart, R. and Baker, W.E. (2000). World value survey: cultural change, and the persistence of traditional values. *American Sociological Review* 65 (1): 19–51.

6 Haerpfer, C.W., Inglehart, R., Moreno, A. et al. (2020). World Values Survey. http://www.worldvaluessurvey.org/WVSDocumentationWV7.jsp.

7 Deci, E.L. and Ryan, R.M. (1985). *Intrinsic Motivation and Self-determination in Human Behavior*. Plenum.

8 Hull, C.L. (1943). *Principles of Behavior: An Introduction to Behavior Theory*. Appleton-Century.

9 Eagleman, D. and Downar, J. (2016). *Brain and Behavior: A Cognitive Neuroscience Perspective*. Oxford University Press.

10 Perel, E. (2019). *Mating in Captivity: Unlocking Erotic Intelligence*. Harper.

11 Laplanche, J. and Pontalis, J.-B. (1988). *The Language of Psychoanalysis*. Routledge.

12 Frankl, V. (1959). *Man's Search for Meaning*. Beacon Press.

13 Herzberg, F. (1959). *The Motivation to Work*. Wiley.

14 Alderfer, C.P. (1972). *Existence, Relatedness, and Growth: Human Needs in Organizational Settings*. Free Press.

15 Thompson, D. (2018). *Hitmaker: How to Succeed in an Age of Distraction*. Penguin Group.

16 McCrae, R.R. and Costa, P.T. (1991). Adding Liebe und Arbeit: the full five-factor model and well-being. *Personality and Social Psychology Bulletin* 17 (2): 227–232.

17 DeYoung, C.G. (2015). Cybernetic big five theory. *Journal of Research in Personality* 56: 33–58.

18 Carver, C.S. and Scheier, M.F. (1990). Origins and functions of positive and negative affect: a control-process view. *Psychological Review* 97 (1): 19–35.

19 Higgins, E.T. (1997). Beyond pleasure and pain. *American Psychologist* 52 (12): 1280–1300.

20 Emmons, R.A. (1986). Personal strivings: an approach to personality and subjective well-being. *Journal of Personality and Social Psychology* 51 (5): 1058–1068.

21 Kilbourne, J. (1999). *Can't Buy My Love: How Advertising Changes the Way we Think and Feel*. Simon and Schuster.

22 Blank, S. (2014). *Holding a Cat by the Tail*. K&S Ranch.

23 Gnyawali, D. and Park, R. (2011). Co-opetition between giants: collaboration with competitors for technological innovation. *Research Policy* 40 (5): 650–663.

24 Elkington, J. (1998). *Cannibals with Forks: The triple Bottom Line of 21st Century Business*. Capstone Publishing.

25 Dyllick, T. and Hockerts, K. (2002). Beyond the business case for corporate sustainability. *Business Strategy and the Environment* 11 (2): 130–141.

4

A Unified Approach to Innovation Marketing with Applications to Organization Readiness in Intrapreneurship Management

Based on the joint investment perspective for innovation in Chapter 2, successful innovation requires the alignment of all participants: Innovators, Supporters, and Adopters, of their respective perceived expected values and risks in investing time, money, effort, and other resources to implement the products (goods or services) developed from an idea to achieve significant impact from its final adoption. Marketing is the process of affecting such alignment by convincing all participants that the innovative product meets the value and risk criteria for their individual investments and committing them to implementation.

Historically, marketing has focused mainly on Adopters. However, innovation marketing needs to focus on all participants. This focus is particularly important for Intrapreneurship, or Internal Innovation in an organization, because of the existing close relationship between Intrapreneur and Internal Supporter. Moreover, based on the new perspective of human needs and wants developed and presented in Chapter 3, these innovation values, and risks (of not achieving the values) originate from the fulfillment of a mixture of human physical and psychological needs for security and wants for stimulation, and evolves with a changing business environment.

This chapter starts with a broad definition and a supply and demand perspective of innovation marketing. It next presents a unified approach to innovation marketing and an integrated system framework for assessing the conditions of market entry for an innovative product from the perspectives of not only Adopter but also providers of innovation, i.e., Innovator and Supporter that provide the innovative product, based on their evolving needs and wants in the business environment. It then applies this unified approach to Intrapreneurship as a basis for both developing an innovative organizational culture and assessing the Organization Readiness in Intrapreneurship Management. This unified approach will be further applied in the next chapter to Adopter as the basis for assessing the

Intrapreneurship Management: Concepts, Methods, and Software for Managing Technological Innovation in Organizations, First Edition. Rainer Hasenauer and Oliver Yu.

Published 2024 by John Wiley & Sons, Inc.

Market Readiness in Intrapreneurship Management. However, it should be emphasized that the purpose of these chapters is not to present any new theories for the well-developed field of marketing but a unified approach and integrated framework of innovation marketing for Intrapreneurship Management. Finally, the chapter discusses the three key elements of Organization Readiness: Innovative Culture, Internal Support, and Innovation Team, and presents practical concepts and proven tools to be used for developing checklists and questionnaires for assessing Organization Readiness.

4.1 A Broad Definition and a Supply and Demand Perspective of Innovation Marketing

As a well-developed field, marketing has many definitions dependent on applications and contexts [1]. However, based on the simple definition of innovation and the joint investment perspective for innovation participants, we may define innovation marketing broadly as "the communication process to attract, engage, and convince an intended innovation participant to accept the perceived expected value and risk of an innovation in satisfying its physical and psychological needs and wants and to commit to investing in the implementation process." With this broad definition, innovation marketing by an organization needs to focus not only on the Adopter of innovation but also all other participants in the Internal Innovation process. This broadened definition is useful for avoiding the frequent overemphasis on a particular innovation participant, such as Final Adopter ("Customer is always right"), Internal Supporter ("We serve only the interest of the shareholder"), or Intrapreneur ("The happiness of the employees is the most important consideration of a business"), since all participants are needed for their joint investment in innovation implementation. Moreover, it leads to not only an emphasis on the communication nature of marketing but also a supply and demand perspective in that the marketer wants to market the supply of an innovation to meet the demand for innovation by the marketee.

This supply and demand perspective is well-accepted when an organization markets the supply of an innovative product to meet the demand for the product by Adopter, where the price is the result of a balance between the cost and desired profit for the product by the supplying organization and the perceived value of the product determined by Perceived Usefulness (PU) and perceived Ease of Use of the innovative product demanded by Adopter. However, this perspective also applies to the mutual marketing between Innovator and the Supporter in general, and to the internal mutual marketing between Intrapreneur and Internal Supporter in an organization as a key element of Intrapreneurship Management in particular. Specifically, with this perspective, the Intrapreneur wants to market the supply of an innovative idea and the related product development plan to meet the Internal Supporter's demand for innovation, and the price is the resulting balance between the resources required together with the physical/psychological rewards desired by Intrapreneur and the perceived value of the innovation development to the organization by Internal Supporter. Moreover, this internal marketing goes both ways, as Internal Supporter also wants to market the supply of an innovative organizational culture to meet the demand for such a culture by Intrapreneur, and the price is the result of a balance between the financial and management resources invested by Internal Supporter for developing such an Innovative Culture and the energy and time required for actively participating in the culture by the Intrapreneur. This internal mutual marketing between Intrapreneur and Internal Supporter is the basis for developing an Innovative Culture and effective Intrapreneurship Management for an organization. This supply and demand perspective of innovation marketing both inside and outside an organization is summarized in Table 4.1.

4.2 A Unified Approach to Innovation Marketing

This broadened definition and the supply and demand perspective further provide the basis for a unified approach to innovation marketing, which aims to persuade and convince all innovation participants to align their investment perspectives

Table 4.1 Supply and Demand Perspective of Innovation Marketing for Intrapreneurship

From \ To	Intrapreneur	Internal Supporter	Adopter
Intrapreneur	x	Supply of innovative idea/product that adds value to the organization. Demand for innovative culture and support that empowers Intrapreneur.	Supply of innovative product that adds value to Adopter
Internal Supporter	Demand for innovative idea/product that adds value to the organization. Supply of innovative culture that empowers Intrapreneur.	x	
Adopter	Demand for innovative product that adds value to Adopter		x

and commit to the effective development and implementation of an innovative product to achieve significant impact. This unified approach to innovation marketing has the following special features.

4.2.1 A Unified Set of Stages for Innovation Marketing with Special Applications to Intrapreneurship Management

With the unified approach, innovation marketing consists of the following five stages.

Stage 1. Attract the initial attention of the intended innovation participant

In the first stage, the marketer needs to attract the attention of intended innovation participant as the marketee. Specifically,

- Innovators may be attracted by the inspiring vision of the innovation in satisfying its needs and wants for exciting meaningful work or financial reward.
- Supporters may be attracted by the potential of the innovation in satisfying its needs for financial benefits, customer appreciation, competitive advantages, or societal rewards.
- Adopters may be attracted by the anticipated benefits of an innovation, including reduced cost, improved quality, exciting aesthetics, prestige, and environmental or societal values in satisfying its needs and wants for various physical and psychological security and stimulation.

For Intrapreneurship Management, because of the special organizational relationship between Intrapreneur and Internal Supporter, a well-developed and actively maintained Innovative Culture is essential to make developing and supporting innovation attractive to Intrapreneur and Internal Supporter, respectively, which can also make the organization and its products appealing to Adopter.

Stage 2. Engage the intended innovation participant in continuing interactions

After initial attraction, the marketer needs to engage the intended innovation participants as marketee in continuing interactions to satisfy the marketee's psychological security needs for building rapport and trust with the marketer so that the marketer can gain further understanding of the perceived values and risks by the marketee for the investment in innovation implementation. At the engage stage, it is vitally important for the marketer to share with the marketee the exciting vision of the innovation development and implementation and the anticipated value impacts in terms of expected economic gains, perceived ecologic benefits, and potential contributions to social equity. Specifically,

- Innovator may be engaged by the psychological stimulation of developing a practical plan for converting an idea into a marketable product with significant values.

- Supporter may be engaged by the psychological stimulation of building a workable process to support the Innovator in developing, producing, and marketing the innovative product.
- Adopter may be engaged by the psychological stimulation of acquiring additional knowledge of the prospective innovative product.

For Intrapreneurship Management, because of their special relationship, successful engagement of Intrapreneur and Internal Supporter will require not only mutual trust but also genuine appreciation, authentic communications, and respectful interactions in developing a common vision of the innovative product development and implementation.

Stage 3. Convince the intended innovation participant to accept the perceived expected values and risks of the innovation

Following engagement, the marketer needs to convince the intended participant as marketee to accept the expected values and risks of the innovation. Specifically,

- Innovator may be convinced by a rigorous value and risk assessment for participating in the development of the innovation, including the availability of risk-mitigating support.
- Supporter may be convinced by an in-depth value and risk assessment of both the creative idea and the quality of the Innovator for developing and implementing the innovation.
- Adopter may be convinced by a credible value and risk assessment for adopting the innovation through the willingness to pay and to use the resulting product.

For Intrapreneurship Management, because Internal Supporter often has many competing values, such as concentrating resources to expand existing successes versus diverting resources to develop new innovations, it is essential for Intrapreneur and Internal Supporter to work closely in iterative market and technology assessment of the expected value and risk of innovation development and implementation. Here, inputs and feedback from Adopter are critical for the assessment to be convincing to both Intrapreneur and Internal Supporter. The Market and Technology Readiness analysis in Chapters 5 and 6 will be particularly relevant for this stage.

Stage 4. Commit the intended innovation participant through negotiation and agreement to invest in the implementation of the creative idea

The proof of success for stage 3 is in the commitment of resources by the intended participant. Specifically,

- Innovator may commit by negotiating an agreement to satisfy its needs for trust and equity in investing in initiating a creative idea and developing product from the idea.

- Supporter may commit by negotiating an agreement to assure that the innovation will satisfy its investment value and risk needs.
- Adopter may commit by negotiating a purchase agreement on the price and performance of the innovation to satisfy its needs for trust and equity in expected value received from investing in the purchase and use of the innovative product.

For Intrapreneurship Management, this commitment is secured in the form of a Champion for the innovation project which can be either Intrapreneur or Internal Supporter. A Champion is a person who has been attracted to, is already engaged in, and has become so convinced by the expected value and risk of the project that he/she is willing to commit all available resources to make it a significant impactful success. In addition to its own commitment to effectively mitigate risks and achieve values of the innovation project, a Champion also needs to secure commitment of the organization by facilitating the understanding and agreement of senior management on the importance of the innovative product to the long-term future of the organization, coordinating needed expertise from other parts of the organization, and protecting Innovation Team from the competing resource demand from the regular production and operations of the organization.

Stage 5. Maintain relationship with intended innovation participant for continued interactions

Because innovation marketing is a continuing interactive process, following the commitment to invest in the innovation project by the intended participant, the marketer needs to continue:

- Maintaining a fulfilling work environment for the Intrapreneur.
- Maintaining fulfillment of the agreement for the Internal Supporter.
- Maintaining warranty and follow-up services for the Adopter, which extends the duration of the impact of the innovation.

All these efforts are for the purpose of satisfying the needs of individual participants for continued trust in each other, for investing in future innovations, and for achieving value at acceptable risk.

Based on the above discussion, the representative needs and wants of the innovation participants to be satisfied for innovation marketing for an organization are summarized in Table 4.2.

For Intrapreneurship, this unified approach highlights the importance of mutual marketing between Intrapreneur and the organization, as represented by Internal Supporter; while Intrapreneur needs to attract, engage, convince, commit, and maintain support from the organization, the organization also needs to attract, engage, convince, commit, and maintain the Intrapreneur for continuous Internal Innovation. To start, for an organization to attract Intrapreneurs, it must

Table 4.2 Representative Needs/Wants Satisfied by the Participants' Perceived Expected Values and Risks of Investing in Innovation Implementation at Different Stages of Innovation Marketing

Marketing Stage	Intrapreneur	Internal Supporter	Adopter
Attract	Inspiring vision, meaningful work, Innovative Culture, financial rewards, organization appreciation.	Financial returns, customer appreciation, competitive edge, societal approval, organization appreciation.	Reduced cost, improved quality, exciting aesthetics, prestige as Innovation Lead, environmental or societal values.
Engage	Psychological stimulation from problem-solving and value creation.	Psychological stimulation from providing support to innovation and value creation.	Psychological excitement in anticipating an innovative product.
Convince	Rigorous value and risk assessment of the innovative product and risk mitigation from internal support.	Comprehensive and in-depth value and risk assessment of innovation development and implementation plan.	Credible value and risk assessment of the innovative product regarding willingness to pay and to use the product.
Commit	Agreement that satisfies the needs for trust and equity in investing in innovation development and implementation.	Agreement that assures the innovation will satisfy the need for trust and benefit to satisfy the corporate value and risk need.	Agreement on performance and price of innovative product to satisfy needs for trust and equity in expected value received from investing in the purchase and use of the innovative product.
Maintain	Maintaining a fulfilling work environment.	Continuing mutually beneficial agreement with Innovator.	Maintaining warranty and follow-up services.

have a proven image backed by a genuine, reliable, and effective Innovative Culture. Effective innovation culture supports Internal Innovation marketing as a multistage and multi-feedback loop process. These loops are essential for generating and sustaining mutual trust among all participants in innovation development and implementation. As to be discussed next, such mutual trust is created by open communication among all participants from different professional disciplines. Multidisciplinary communication integrating a wide range of domain expertise requires a set of well-defined communication rules and sincere mutual respect and tolerance in innovation development and implementation under tight time-to-market schedules and short Innovation Half-Life in the face of strong competitors to Innovation Team, which will be discussed further in Chapter 5.

4.2.2 Major Challenges and Solutions for Innovation Marketing

This unified approach identifies not only the key elements but also the major challenges for innovation marketing. For a marketer to effectively attract, engage, and convince an intended innovation participant as a marketee to commit to investing in innovation implementation first requires mutual rapport and trust. To develop true rapport, the marketer must have genuine respect and empathy for the needs and wants of the marketee. To achieve real trust, the marketer must have the integrity to be truthful with and commit to its promise to the marketee. These requirements are difficult because of the often conflicting interests between the marketer and the marketee. Specifically, the marketer is tempted to obtain the commitment of the marketee not to satisfy the needs and wants of the marketee but those based on the marketer's own self-interest and even for ego gratification. As a result, the marketer would be tempted to view the marketee as intellectually inferior and often use exaggerated and even false promises to deceive the marketee, which will eventually totally destroy mutual trust and long-term relationships with each other.

The mutual marketing between Intrapreneurs and Internal Supporter has its own special challenges. First, there are many conflicting interests, such as the preoccupation with existing successes, the pressures from shareholders for short-term profits, and the fear of failure from new ventures for the potential Internal Supporter in the organization. Then, there are many complex resources and other requirements, such as sufficient financial and personnel commitments, effective communications about policies and procedures, and productive incentives and training programs.

To effectively meet these challenges, a review of major research and case studies [2] indicates the following general elements for the solution:

- A marketer must recognize that marketing is a long-term engagement and the marketee is an intellectual equal who can be a productive collaborator if treated with respect, rapport, integrity, and trust.
- A marketer must subscribe to the basic ethics of honesty and integrity in its interactions with the marketee as the foundation of all marketing activities.
- A marketer must develop strong communication skills, including abilities to build a trustworthy reputation and image, gauge accurately the marketee's values and emotions, use language and media proficiently, develop close rapport with marketee, and form strong allies with powerful influencers to amplify the communication message.

- A marketer needs to recognize genuine differences in the perspectives as well as limitations in receiving and utilizing marketing information from the marketee. These differences and limitations are particularly important among marketees with different personalities, cultural backgrounds, intellectual capabilities, and professional disciplines. This element is particularly important in marketing to Adopter and will be discussed further in Chapter 5.
- For Intrapreneurship, an organization must develop and maintain an effective Innovative Culture that explicitly defines its vision and values, encourages fast learning from experimentation, and rewards creativity. This critical element will be discussed in-depth in the next section including summaries of case studies of successful culture development in several major innovative organizations.
- For Intrapreneurship, an organization must also actively develop effective Innovation Team building capabilities through training and practices, as innovation is by nature a team sport that requires the participation of diverse individuals in different roles and functions, including the Champion, the idea generator, the project planner, the product developer, and marketer, together with Internal Supporter. This important element will be further discussed in the next sections.

4.3 Application of the Unified Approach to Innovation Marketing in Intrapreneurship: Organization Readiness Assessment

Since Intrapreneurship requires mutual marketing between Intrapreneurs and Internal Supporter, the unified approach to innovation marketing can be applied to the development of Organization Readiness for innovation with key elements shown in Figure 4.1, with RL for Readiness Level.

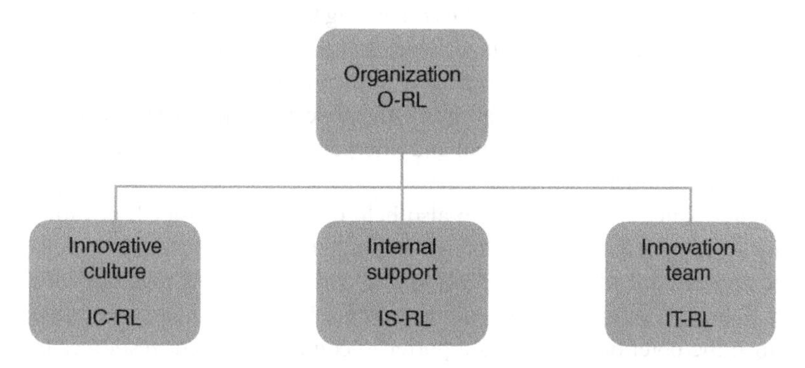

Figure 4.1 Key Elements of Organization Readiness for Intrapreneurship

Specifically, for an organization to be ready for Internal Innovation development, it must have a well-established Innovative Culture, a fully developed internal support system, and a highly effective Innovation Team. These key elements will be further discussed below.

4.3.1 Innovative Culture

To *attract* and *engage* potential Intrapreneurs, an organization needs to have an Innovative Culture that serves as an advertisement. Organizational culture is a highly developed field of study. However, for Intrapreneurship Management, we may define it as "a set of explicit and implicit operational and interpersonal rules and code of conduct collectively perceived by the employees of an organization as the basis for their behavioral strategies to survive and thrive in the organization." (In a broader context, regional or national culture may be similarly defined as a set of explicit and implicit operational and interpersonal rules and code of conduct collectively perceived by the residents of a region or nation as the basis for their behavioral strategies to survive and thrive in the region or nation.) Because the senior management of an organization is responsible for the explicit operational rules and code of conduct, organizational culture development necessarily starts from the top.

Based on this definition, there are two major sources of organizational culture: one is the explicitly stated expectations and code of conduct for employee behavior, which is generally proclaimed by the top management of the organization and can certainly influence employee behavior; and the other is the perceived actual practice or tolerance of various employee behavior, which generally has a much more powerful influence.

To specifically develop an Innovative Culture, the top management of an organization must personally market to employees the ethical, clear, and effective communications of the common values about innovation. The ethics, clarity, and effectiveness of communications must be convincing to and verifiable by employees through observable commitments to clearly stated common values and goals for innovation, fair and equitable incentives, well-coordinated support, rigorous assessment, and visible tolerance of failures. Otherwise, the employees will necessarily form an implicit set of operational and interpersonal rules and code of conduct as a counter-productive culture.

On the other hand, an employee can also help develop and foster Innovative Culture in an organization with effective internal marketing through a well-researched and carefully developed idea and implementation plan, together with the ability to build a competent and committed core Innovation Team to engage and convince the potential Internal Supporter to commit sufficient resources to attain success that would, in turn, reinforce the credibility of the innovativeness of the employee and grow the Innovation Team to transform the organizational culture.

4.3.2 Internal Support

To *convince* and *commit* potential Intrapreneur to Internal Innovation, an organization must demonstrate its own commitment to Innovative Culture through an active and organized internal support system, which includes concrete incentives and channels for promoting idea generation and business plan development; rigorous review and assessment of business plans; active guidance, mentoring, and funding for product development and marketing; and other management and administrative support.

The case studies of several major innovative organizations are presented in Appendix A and summarized below to provide concrete examples of the Internal Support to Intrapreneurs.

- *3M*: The senior management of 3M has emphasized the importance of innovation from its inception in 1902 through the formation of internal Tech Forum, major awards for innovators, and the 15% culture that allows each employee to use 15% of their working time to pursue projects of personal interest, which started in the 1950s, decades before other adoption by other innovative companies like Google.
- *IBM*: For another over a century-old company, IBM has long fostered the innovation culture through the highly respected and rewarded technical career track. In addition, it has made ethics not only a formal corporate value but also a strictly enforced practice.
- *SRI International*: SRI was formed in 1946 by Stanford University to conduct cutting-edge research with over 3,000 major innovations including the Mouse, Window technology, and Siri. One of its technical staff members, Dr. Curtis Carlson, rose through the ranks to become the longest-serving (16 years) President and CEO and instituted the Value Creation Forum and other tools to support continuing innovation success. These tools formed the basis of a mandatory training program for all new employees at SRI and a widespread practice for all technical staff. Moreover, Dr. Carlson used his regular personal participation in the program to not only energize employees with senior management attention and commitment, but also provide vital support to innovators with rigorous assessment, effective mentoring and supervision, and fast funding process.

4.3.3 Innovation Team

The third key element of Organization Readiness for Internal Innovation is the Innovation Team Readiness. The unified approach to innovation marketing can again be applied to assess this special element of Organization Readiness. There are two aspects to this application: First, there is the internal marketing within the team. Specifically, the initial idea generator or Adopter of an existing idea needs to market to a core team of diverse domain experts, including engineering designer, customer behavior analyst, market strategist, legal adviser, and production specialist, to attract, engage, convince, and commit them to jointly invest in the development and implementation of the product derived from the creative idea. Second, because different from a well-defined project like building construction, an innovation project deals with mainly new and mostly untried ideas and concepts that need frequent modifications. As a result, the traditional command-and-control step-by-step approach for team and project management is not compatible with the management of an Innovation Team and project. This fundamental difference from the traditional approach must be fully recognized and understood by both Innovation Team and Internal Supporter.

This fundamental difference also indicates that an effective Innovation Team must not only have diverse domain expertise but also complementary personality traits. Furthermore, innovative product development must be flexible and continually evolving. Fulfilling these special characteristics of team and product development and management requires effective team building, which is essential for an innovative organizational culture.

We will present here two different approaches to team building based on proven successful concepts and practices which are basically consistent with the unified approach to mutual marketing among team leaders and team members by attracting, engaging, convincing, committing, and maintaining collaborative relationships with each other through an understanding of common needs and wants and building mutual trust for productive interactions.

4.3.3.1 The Agile and Scrum Approach to Team Building

This approach was originally developed for computer software project management because of the constantly evolving nature of software development. However, since software is a special example of innovative products, the well-known values and principles of Agile and Scrum [4, 5] are fully applicable to all Innovation Teams and product development and management by simply replacing "software" with "innovative product" as shown below.

Four Team Values:

- Individuals and interactions over processes and tools.
- Working on *innovative products* over comprehensive documentation.

- Customer collaboration over contract negotiation.
- Responding to change over following a plan.

12 Team Principles:

1) Our highest priority is to satisfy the customer through early and continuous delivery of valuable *innovative products.*
2) Welcome changing requirements, even late in development. Agile processes harness change for the customer's competitive advantage.
3) Deliver *innovative products* frequently, from a couple of weeks to a couple of months, with a preference for the shorter timescale.
4) Businesspeople and developers must work together daily throughout the project.
5) Build projects around motivated individuals. Give them the environment and support they need and trust them to get the job done.
6) The most efficient and effective method of conveying information to and within a development team is face-to-face conversation.
7) Working on *innovative products* is the primary measure of progress.
8) Agile processes promote sustainable development. The sponsors, developers, and users should be able to maintain a constant pace indefinitely.
9) Continuous attention to technical excellence and good design enhances agility.
10) Simplicity – the art of maximizing the amount of work not done – is essential.
11) The best architectures, requirements, and designs emerge from self-organizing teams.
12) At regular intervals, the team reflects on how to become more effective, then tunes and adjusts its behavior accordingly.

Applying this equivalence relationship also yields the following application of Scrum concept to Innovation Team and Internal Supporter in Intrapreneurship Management (Table 4.3).

The official Scrum roles are Product Owner, Scrum Master, and Developer. For Intrapreneurship development and management, the Product Owner is the Innovation Team Champion, who could be the idea generator or Adapter, another fully dedicated member of the Innovation Team, or even the Internal Supporter, defines the attributes of the product, agrees, and complies with the requirements of the Steering Committee organized by Internal Supporter as well as with the demand of the internal and external Adopters.

The Scrum Master is the Internal Supporter, who surveys the Agile process, removes obstacles, provides optimal working conditions, and responds to the requirements of the Steering Committee.

The Developer is the Innovation Team, which develops the product in compliance with the Product Owner's targets and includes functional resources such as quality assurance and productivity experts as well as market experts.

Table 4.3 Internal Supply and Demand in Intrapreneurship Management

Scrum Roles	Tasks	Intrapreneurship Equivalent Roles
Product owner	Responsible for the increase of business value of the *innovative product*. Defines innovation content, and reports to the Steering Committee, in compliance with functional roles of *Internal Supporter and Adopter*	Champion of the Innovation Team, who could be idea generator or Adopter, or another fully dedicated member of the team, or even the Internal Supporter
Scrum master	Supports *Innovation Team* to optimal use of Scrum	Internal Supporter (Enabler)
Developer	Develops the *innovative product*	Innovation Team
supporter	Comprises all indirectly engaged assistive experts	Internal Supporter
Steering committee	Decides on budgetary allocation, releases for next development stage	Internal Supporter organized Steering Committee
end user, customer, stakeholder	Decides on usability and applicability of *innovative product*	Internal and External Adopters

The relationships and interactions among Innovation Team (Idea Generator and Developer), Internal Supporter, and Final Adopter (individual customer or Adoption Committee) can be further illustrated based on the Scrum concept in Figure 4.2.

From these relationships and interactions, we can again adapt the Scrum concept to develop the following special assessment of the Organization Readiness of an Innovation Team and product development and management:

1) Are the roles of the team members well-defined and complementary?
2) Is customer satisfaction the utmost goal of product development?
3) Has there been constant customer collaboration?
4) Are the people more valued than the product?
5) Is the flexibility of the product development fully appreciated and supported by the organization?
6) Have there been frequent meetings among team members and with customers?
7) Have the businesspeople always been working with product developers?
8) Has there been continuous development of trial products?
9) Is simplicity always valued?
10) Is the team self-organizing?
11) Is technical excellence used for agility?
12) Has there been regular reviews?

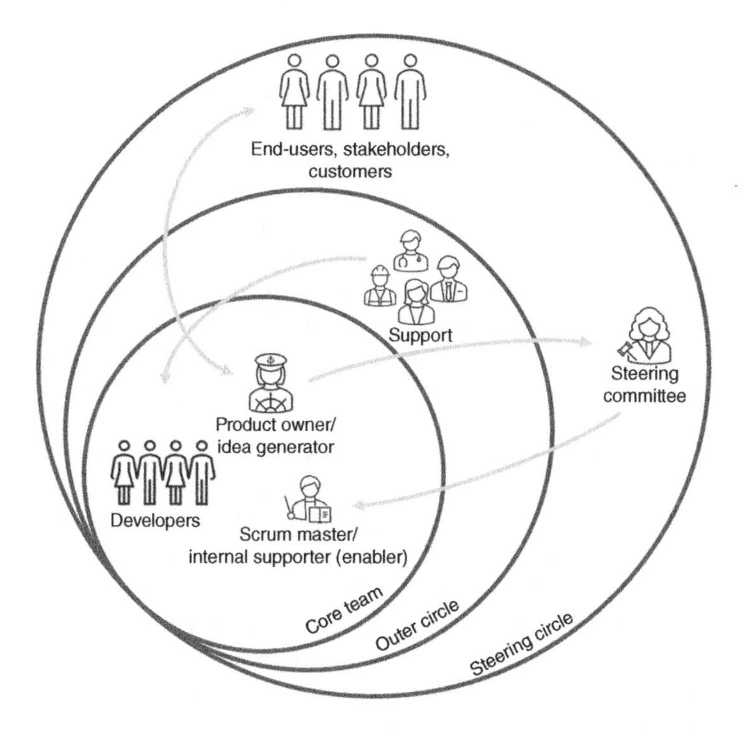

Figure 4.2 Relationships and Interactions of Innovation Team Adapted from Scrum

4.3.3.2 A Novel Approach for Applying the Enneagram to Innovation Team Building (Based on contribution by M. Schlegel)

The Enneagram is an ancient concept of human dynamics with roots tracing back thousands of years. It has been popular in the modern form as a personality assessment system since the early 20th century. At the beginning of this century, professional engineering trainer, Matthew Schlegel, recognized a close correspondence between the personality types and dynamics of the Enneagram and those involved in Innovation Team building and problem-solving. Through extensive research and proven application successes, Schlegel developed a novel approach for applying the Enneagram to Innovation Team building and creative problem-solving [6]. The application to Innovation Team building will be presented here, and the application to creative problem-solving will be presented in Chapter 6. It should be noted that although it is possible to apply similar personality types and dynamics from other systems like the Big Five/OCEAN or Myers-Briggs, it is the remarkable close correspondence between those in the Enneagram and those in Innovation Team building and problem-solving that make the Enneagram a particularly good fit for the application.

- *Overview of the Enneagram Personality System*

The word *Enneagram* is composed of two parts, *Ennea* meaning nine in Greek and *gram* meaning figure. The Enneagram symbol is a nine-point diagram for a personality dynamics system shown in Figure 4.3.

The Enneagram is organized into main groups of three. These groups of three, or triads, are called the *Main Triads* or *Centers*. Associated with each one of these centers is a body part, a body activity, and an emotion. Types 8, 9, and 1 are in the *Gut* or *Intuitive Center*. These types start with their intuition and are strongly under the influence of anger. Types 2, 3, and 4 are in the *Heart* or *Feeling Center*, start with their feelings, and are influenced by all emotions. Types 5, 6, and 7 are in the *Head* or *Thinking Center*, start with thinking, and are strongly influenced by anxiety.

A neurological basis has been hypothesized for the nine types related to the two parts of the brain that modulate the behaviors described by the nine Enneagram types: the amygdala and the pre-frontal cortex (PFC). In the bilateral human brain, there exist two amygdalae and two PFCs, a right and left of each. In the same manner that asymmetrical dominance of the motor cortex modulates the three states of handedness – left-handed, right-handed, and ambidextrous – the amygdalae and the PFCs each have three dominant modes – left dominant, right dominant, and ambiguated. From the three-by-three combinations of these modes arise the nine distinct Enneagram types.

Within each center, there is another layer of organization according to the relationship of the type with its dominant emotion: externalized expression, internalized expression, or suppressed expression. For the Gut types, the 8 externalizes

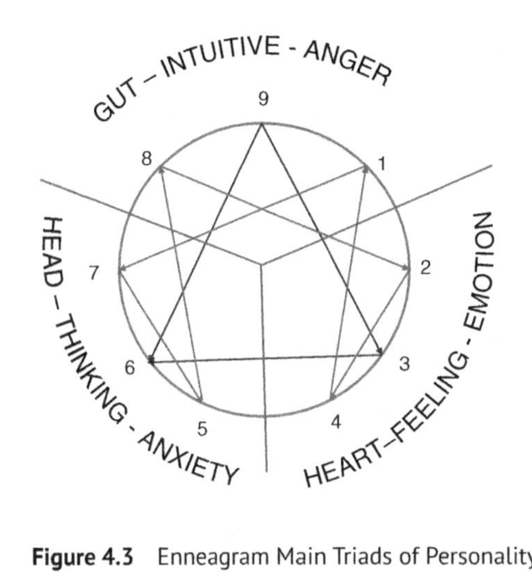

Figure 4.3 Enneagram Main Triads of Personality Dynamics in an Innovation Team

anger, the one internalizes anger, and the 9 suppresses anger. For the Heart types, the 2 externalize emotions, the 4 internalize them, and the 3 suppress emotions. For the Head types, the seven externalizes anxiety, the 5 internalizes anxiety, and the 6 suppresses it. The *suppressed types* are also called the *core types*. The core types are connected by the arrows forming a triangular figure of 3, 6, and 9. The other types are connected to form an irregular hexagonal figure of 1, 4, 2, 8, 5, and 7.

There have been varying but basically, similar labels and motivations or personality dynamics ascribed to the Enneagram types. Table 4.4 combines the inputs from several sources [6].

There are several ways to have highly accurate tests to determine a person's starting or dominant Enneagram type:

1) Self-assessment test
2) One-on-one test with an Enneagram facilitator
3) Group setting with an Enneagram facilitator

Each way has its pros and cons depending on the comfort level of each type of person. For example, a Type 5, who is uncomfortable disclosing personal information to others, will prefer option (1). More social types may prefer option (2), while still others would prefer option (3).

In addition to personality type assessment, the Enneagram is by design a dynamic system that describes how each type's behaviors vary based on their level of confidence or security. The lines within the Enneagram indicate these dynamics. It is important to know that a person can have inherently or learn to acquire more than

Table 4.4 Enneagram Type, Label, and Motivation or Personality Dynamics

Enneagram Type	Label	Motivation or Personality Dynamics
1	Perfectionist	Determine how things should be and right the wrongs
2	Helper	Help others to receive appreciation
3	Achiever	Actively contribute to recognition by others
4	Individualist	Feel all emotions and communicate evocatively
5	Investigator	Collect information and conduct analysis to gain understanding e
6	Planner	Anticipate future to reduce risks
7	Enthusiast	Promote fun to be liked by others
8	Protector	Secure control of the environment
9	Mediator	Minimize interpersonal conflicts

one personality type. Typically, people have a starting type and can access the behavioral dynamics of other types through the stress lines in Enneagram. For example, a Type 1 will have access to the behaviors associated with their stress line connection to types 7 and 4. Here, the direction of the arrows in these lines is important as they indicate positive reinforcements. For example, when Type 1 is feeling confident or secure, it moves in the direction of the arrow toward Type 7 and starts to exhibit Type 7 behavioral dynamics. When Type 1 feels less secure, it moves backward in the direction of Type 4 exhibiting Type 4 behaviors. Additionally, each type can access some of the dynamics of the type on either side. These adjacent types are called the *wing types*. For example, for Type 1, the wing types are 9 and 2.

- *Matching Enneagram types to Problem-Solving Steps in Innovation Development*

Based on both analysis and experience, Schlegel found [6] that a systematic creative problem-solving process contains the logical steps shown in Table 4.5.

By comparing Tables 4.4 and 4.5, Schlegel found a perfect match between the Enneagram types and the personality types needed to lead the team in different steps of problem-solving. The comparison can be directly seen from Table 4.6 which combines Tables 4.4 and 4.5 side-by-side.

With successful applications of matching people with the corresponding personality types to the problem-solving steps, Schlegel has concluded that a systematic approach to problem-solving, intentionally taking a team through each of the nine dynamics, yields successful and resilient resolution of the original problem.

It would be ideal if the team could have all nine Enneagram personality types; however, that is not necessary because a person can access several personality types through stress line connections and wing types. In fact, if the personality

Table 4.5 Logical Steps of a Systematic Creative Problem-Solving Process

Step	Description
1 – Problem-goal	Identify the problems and define the goals
2 – Stakeholders	Recruit a committed team
3 – Ideation	Generate ideas for solutions
4 – Emotional reactions	Assess reactions to each idea
5 – Logical analysis	Study and score promising ideas
6 – Planning	Select the most promising ideas and build an action plan
7 – Promotion	Passionately promote the plan and get approval to proceed
8 – Implementation	Execute the plan and solve the problem
9 – Integration	Confirm that the problem is solved with stakeholders

Source: Schlegel, Teamwork 9.0.

Table 4.6 Comparison of Tables 4.4 and 4.5

Step	Description	Type & Label	Personality Dynamics
1) Problem-Goal	Identify problems and define goals	1) Perfectionist	Determine how things should be and right the wrongs
2) Stakeholders	Recruit a committed team	2) Helper	Help others to receive appreciation
3) Ideation	Generate ideas for solutions	3) Achiever	Actively contribute for recognition by others
4) Emotional Reactions	Assess reactions to each idea	4) Individualist	Feel all emotions and communicate evocatively
5) Logical Analysis	Study and score ideas	5) Investigator	Collect information and conduct analysis to make sure
6) Planning	Select best idea and build an action plan	6) Planner	Anticipate future to reduce risks
7) Promotion	Promote plan for approval	7) Enthusiast	Promote fun to be liked by others
8) Implementation	Execute plan to solve problem	8) Protector	Secure control of the environment
9) Integration	Confirm problem solved with stakeholders	9) Mediator	Minimize interpersonal conflicts

dynamics of each type are well-understood, even a single individual can use them to perform the corresponding roles for successful problem-solving.

- *Application of Enneagram Triads to Innovation Team Building*

There are certain Enneagram types that have an affinity for one another. One such set of grouping was identified by the noted Stanford medical professor David Daniels [7] as Harmony triads, which are formed by the following types: Types 1, 4, and 7; types 2, 5, and 8; and types 3, 6, and 9. In the applications of Enneagram to Innovation Team building, Schlegel has observed that workers in each Harmony Triad tend to work well with each other and form a sub-team organically. Based on the neurological hypothesis for Enneagram types, the inner harmony of these Harmony Triads may occur due to the perfect balance of the brain asymmetries within each Triad by having all modalities equally represented. This balance makes the types in these groups complementary and neurodiverse.

As discussed previously, each Enneagram type contributes a distinct perspective and motivation. When they form a Triad sub-team, these perspectives and motivations flavor the overall focus of the Triad sub-team and guide its common interests and goals. Schlegel gives these sub-teams names that reflect the focus of each: the Start-Up Triad (types 1, 4, and 7), the Industrious Triad (types 2, 5, and 8), and the Systematizing Triad (types 3, 6, and 9).

With the Start-Up Triad, Type 1 brings a focus to identifying a problem that needs to be solved. Once alerted to the problem, Type 4 will be the first to react emotionally "feel" to the problem. Type 7, wanting to avoid negative feelings associated with a "problem," will be compelled to find solutions for making everyone happy again. Furthermore, Type 1 brings an intense scrutiny to and understanding of the problem. Type 4 will "see" what is missing and present creative ideas. Type 7 will synthesize all ideas, collected both from within the team and without and will propose solutions to the problem. Cycling through problem-solving, Type 1 will assess proposed solutions analytically in terms of their efficacy in resolving the problem, and identifying any weaknesses and deficiencies. Type 4 will assess the proposed solutions emotionally, and how positively the solutions will be received by stakeholders. And Type 7 will promote promising solutions to the broader community. The specific dynamics that these three types deliver to the team are ideally suited to creative problem-solving. Schlegel observed that these dynamics are present in early product development cycles, both in start-up companies and in innovation centers in larger companies. Curiously, once there is a clear path forward to resolving the problem, this team will often turn their attention to a new "fun" problem to solve rather than dwelling on the same problem until it is ultimately solved. At that point, the Start-Up Triad hands the project off to the Industrious Team.

The Industrious Triad, dominated by Type 2, 5, and 8 dynamics, starts with Type 2 in wanting to help further the efforts of others and solve a problem, in this case implementing the solution proposed by the Start-Up team. Type 5 will take the promising idea, analyze it, and determine how to implement it in a way that will actually solve the problem. Type 8 will build prototypes and actively troubleshoot any implementation problems that arise until full-scale implementation is possible. Turning back to the stakeholders, Type 2 will ensure that the implementation meets the objectives of the end customers. Type 5 will continually reveal clever and more efficient ways to deliver the solution. And Type 8 will ramp up production to meet demand. As the name of this Triad implies, the Industrious Triad is about getting to action and getting things done, characteristics of Type 2 and especially Type 8.

Finally, for the Systematizing Triad consisting of Type 3, 6, and 9, the starting point is Type 3. Motivated by the desire to be continuously ever more successful, Type 3 will deliver ideas to systematize the processes that replicate successful outcomes. Type 6 then develops and elaborates the processes and controls for these systems and provides a perspective of minimizing risk, which is very complimentary to Type 3 in that both types seek to avoid failure. Finally, Type 9 will bring the perspective of other stakeholders during system implementation that ensures consensus and compliance. Type 3 will closely scrutinize the efficacy of the systems and in close collaboration with Type 6 seek to optimize the performance of the system in creating successes. All the while Type 9 will be attentive to all stakeholders and ensure that the systems are working for everyone. The Systematizing Triad will seek to systematize the activities associated with dynamics of the Start-Up Triad and the Industrious Triad to build creative and productive organizations that continuously and successfully solve problems for stakeholders.

In summary, unlike other separate personality dynamics systems and problem-solving systems, the Enneagram approach integrates both aspects into one coherent system and shows how human personality dynamics contribute to problem-solving. This powerful framework can help teams better organize and structure Innovation Teams for problem-solving and can provide a track for the team to follow to reach its goal. Finally, the three Triad dynamics – Start-Up, Industrious, and Systematizing – overlap with characteristics of the innovation cycle team dynamics; the Start-Up Triad relates to Innovation Team, the Industrious Triad to the specific Internal Supporter to the Innovation Team, and the Systematizing Team to the overall Organization Management. The clean and clear alignment of these Triad structures and the role they play in innovation and problem-solving in an organization is precisely the expected evidence corroborating the role human dynamics play in the innovation cycle.

Moreover, as a personality dynamic system, the Enneagram helps members of the Innovation Team understand the different personality styles, motivations, behaviors, and interpersonal dynamics of team members as key ingredients to building effective teams. Each Enneagram type brings a distinct set of capabilities to a team. A team with a variety of Enneagram types will have more ability to examine problems from their respective perspectives. The distinct motivation of each type drives those abilities. Knowing and appreciating the nine distinct styles and contributions allows members of the Innovation Team to better appreciate one another and celebrate style diversity on the team, as style-diverse teams are highly effective problem solvers when working together well.

There are also many other benefits for an Innovation Team that knows the Enneagram types of the team members: improved communication, common vocabulary to resolve interpersonal conflicts, and appreciation for team members' innate capabilities. For leaders, the system provides team member motivations, reveals team strengths and gaps, and provides a prescription of how to fill the gaps and create well-rounded, style-diverse teams. The following case study presents one real-life example of how the system was used by a Type 1 leader to resolve an interpersonal conflict on his team and improve team outcomes:

> A Type 6 team member approached a Type 1 leader with a conflict that had occurred between the Type 6 and a Type 2 teammate. The Type 2 had grown

angry with Type 6 and had even physically pushed Type 6. Type 6 was baffled, not understanding why Type 2 was so angry. Digging into the situation, Type 6 revealed that for several weeks Type 2 was picking up printouts from the printer and delivering them in person to Type 6's office. Type 6 felt bad to trouble Type 2 (who was a vice president at the company!) and suggested that Type 2 really did not need to go to the trouble. The Type 1 leader recognized that Type 6 had unwittingly cut off a source of appreciation for Type 2. Knowing that receiving appreciation is a key motivator for Type 2, the Type 1 leader suggested that Type 6, who communicated regularly with customers, pick one important communication every day to ask Type 2 for help crafting the best response to the customer. At first, Type 6 thought this was a waste of time, but tried it as the Type 1 leader suggested. Type 2 was delighted to help Type 6, and Type 6 generously thanked Type 2 each day. The conflict between them immediately evaporated. Type 2 and Type 6 developed a strong working relationship through that daily routine. Type 6 reported that Type 2 provided insights and perspectives that improved his communications. Type 6 learned an important lesson about appreciation. Carrying the lesson forward, when Type 6 visits a customer in person, if the customer offers coffee, Type 6 will always graciously accept, even if he does not want coffee. Just in case the host is a Type 2, Type 6 does not want to start off the meeting by denying a Type 2 the opportunity to receive appreciation.

It is important to note that in building a team for innovation development and implementation in an organization, the team members initially will most likely not have all the complementary personality traits. Then, applying the Enneagram approach can guide the team building process by emphasizing various effective personality traits at various stages of innovation as well as acquiring team members with additional personality traits as the team evolves and expands.

The Enneagram can be used to further address the following two important aspects of team building and operation:

First, the Enneagram approach for team building can be used to help develop mutual trust and effective communication among members of an Innovation Team as many people have hidden agendas that are not automatically revealed during team formation but could later cause misunderstanding, conflicts, and mistrust. It is important in Step 1 of this approach to have all team members get together for an "airing of the grievances." Have everyone explain the problem(s) from their distinct perspective and have everyone else listen attentively. When the team comes to this first meeting, there may have been mistrust, blame, and finger-pointing going on. The approach ensures the team members share whatever problems they perceive, even those that seemingly are caused by others' actions. When this is done, the group experiences a catharsis. Like the tale of the blind men and the elephant, they come to the realization that they share a common problem – the elephant in the room – that they are all experiencing. This common problem becomes the group's common enemy that will become their central focus. They all realize that they share this common enemy and begin to trust one another so that they can work together to eliminate the common enemy. In the second part of Step 1, the group will create a shared vision for how the workplace will be once the enemy is slain and the problem is solved. The group coalesces around the common problem and the shared vision. Doing Step 1 well mitigates misunderstandings, conflicts, and mistrust.

Second, the Enneagram approach can help an Innovation Team adapt to evolving understanding and goals in problem-solving, which often occurs with new and not well-defined or understood problems. Specifically, during the course of a project, new problems will arise: a new team member joins with a novel perspective on the problem, a problem was obscured at the outset of the initiative, etc. For whatever reason, a new problem is uncovered, and the team, including any new team members, will revisit Step 1, review the problem and vision statements, adjust them accordingly, and assess the impact on subsequent steps. Using the stepwise problem-solving framework, this can be done in a systematic way that grounds the activity, keeps team members focused, and enables the team to use the same framework to get the initiative back on track.

Summary

This chapter starts with a broadened definition of innovation marketing to include marketing among all participants of the innovation process. It also provides a supply and demand perspective and applies it to the marketing among these participants. The chapter next identifies the major stages of marketing consisting of the marketer attracting, engaging, convincing, committing, and maintaining the marketee. It then presents a unified approach to innovation marketing which involves marketing among all participants but focuses this chapter specifically on the mutual marketing between Intrapreneur and Internal Supporter as a special element of Intrapreneurship Management. The chapter next discusses the major challenges and solutions to innovation marketing. It further identifies the key elements of Organization Readiness as the readiness of Innovative Culture, Internal Support, and Innovation Team. For the Innovation Team Readiness, it presents two proven approaches to effective team building: the Agile and Scrum approach that emphasizes the shared values and work principles for an effective team for innovation development and implementation, and the novel approach of applying Enneagram to Innovation Team building that emphasizes the complementarity and balance among the personality dynamics of Innovation Team member for effective problem-solving. Discussion of these key elements will be used as the basis to develop a set of checklists and questionnaires in Chapter 7 for the input of the Organization Readiness assessment as the starting part of the Intrapreneurship READINESS navigator (IRN©) software. It is important to note that the discussions on applying the unified marketing approach to develop an innovative organizational culture are necessarily conceptual and the two proven team building approaches are both process-oriented and interactive. They must be effectively learned through expert-facilitated training together with continuous practices to become proficient and adaptive to changing environments and different applications.

Glossary

Agile and Scrum: A proven approach to team building that is basically consistent with the unified approach to mutual marketing among the team leader and the team members.

Enneagram: An ancient concept on human dynamics that has been proven effective in team building and is conceptually consistent with the unified approach to mutual marketing among the leader and members of a team.

Innovation marketing, a broad definition: The communication process to attract, engage, and convince an intended innovation participant to accept the perceived expected value and risk of an innovation in satisfying its physical and psychological needs and wants and to commit to investing in the implementation process.

Organizational culture: A set of explicit and implicit operational and interpersonal rules and code of conduct collectively perceived by the employees of an organization as the basis for their behavioral strategies to survive and thrive in the organization.

Stages of marketing: Marketer to attract, engage, convince, commit, and maintain marketee.

Unified marketing approach: For Intrapreneurship Management, marketing is not limited to the Final Adopter, but extends to all participants of the Internal Innovation process in an organization, including the marketing within an Innovation Team and the mutual marketing between the Innovation Team and Internal Supporter.

Discussions

- Study online, report, and discuss major definitions of marketing.
- Critique the broad definition of Innovation Marketing.

- Discuss the unified approach to Innovation Marketing.
- Amplify the solutions for the challenges of Innovation Marketing.
- Critique the definition of Organizational Culture.
- Assess the innovativeness of your Organizational Culture.
- Further develop the actional steps to improve the innovativeness of your Organizational Culture.
- Develop the actual steps to improve the internal support system for innovation development and implementation in your organization.
- Assess the effectiveness of your Innovation Team.
- Develop the actionable steps to improve the effectiveness of your Innovation Team by applying the Agile and Scrum approach, the Enneagram approach, or other approaches.
- Discuss the personality dynamics of your team members and use the Enneagram to identify their individual types.
- Evaluate the strengths and weaknesses of your team members and how to evolve, expand, or contract your team as appropriate to improve the performance and meet the challenges.

References

1 Cohen, H. (2011). 72 marketing definitions. https://heidicohen.com/marketing-definition (accessed 29 March 2011).

2 Weber, C., Hasenauer, R., and Mayande, N. (2018). Toward a pragmatic theory for managing nescience. *International Journal of Innovation and Technology Management* 15 (5): https://doi.org/10.1142/S0219877018500451.

3 Eljasik-Swoboda, T., Rathgeber, C. and Hasenauer, R. (2019). Assessing technology readiness for artificial intelligence and machine learning based innovations. *Proceedings of the 8th International Conference on Data Science, Technology and Applications* (26–28 July 2019), Prague, Czech Republic, pp. 281–288. ISBN: 978-989-758-377-3.

4 Agile Manifesto (2001). Manifesto for agile software development. http://agilemanifesto.org/.

5 Schwaber, K. and Sutherland, J. (2013). The scrum guide: the definitive guide to scrum: the rules of the game. Scrum.org.

6 Schlegel, M. (2020). *Teamwork 9.0, Successful Workshop Group Problem Solving Using the Enneagram*. Schlegel Publishing.

7 Daniels, D. (2007). The essential enneagram: the definitive personality test and self-discovery guide. drdaviddaniels.com.

5

Applying the Unified Approach of Innovation Marketing to External Adopter for Market Readiness Assessment

In Chapter 4, we have developed a unified approach to innovation marketing and applied it to the organization to assess its readiness for Intrapreneurship, i.e., internal innovation development. In Chapters 5 and 6, we will apply the approach to assess respectively the Market- and Technology-Readiness of prospective innovative products (goods or services) developed by the Intrapreneur as Innovation Team. Specifically, in this chapter, we will first discuss the product-specific and interactive nature of the general concepts of Market Readiness and Technology Readiness for products (goods or services) to be introduced to the Adopter in the market. We will then examine the evolving marketing targets for innovation, advancing marketing tools, major challenges to innovation marketing, and the special importance of marketing ethics to individual adopters. We will finally develop the key elements and detailed concepts for assessing the Market Readiness of a product as the basis for the input to the Intrapreneurship Readiness Navigator (IRN©) software.

5.1 Interaction Between Market Readiness and Technology Readiness

Different from Organization Readiness, which assesses the general readiness of an organization in developing Intrapreneur and implementing internal innovation, Market Readiness and Technology Readiness are specific to the product developed in response to the needs and wants of the Final Adopter [1, p. 7]. In business-to-consumer (B2C) marketing, the Final Adopter is a private household, a family member, or a private person. In business-to-business (B2B) marketing the Final Adopter is a business unit, a legal unit, a person, or even a

Intrapreneurship Management: Concepts, Methods, and Software for Managing Technological Innovation in Organizations, First Edition. Rainer Hasenauer and Oliver Yu.

Published 2024 by John Wiley & Sons, Inc.

robot, but generally a system. The conceptual interactions between Market Readiness and Technology Readiness of a product are interactive as shown in Figure 5.1.

An innovative product is ready for success when both its Technology and Market Readiness levels reach the "sweet spot," where the market is ready to adopt innovation and technology is ready to fulfill the market need. This is also represented by the "Window of Opportunity" to make a profit, as defined by Day and Freeman [2], the goal of an innovative product.

Additionally, the interactions between the nominal 1–10 scales of Market Readiness Level (MRL) and Technology Readiness Level (TRL) can be used to illustrate the potential market and technology risks of innovation development and implementation as shown in Table 5.1.

This table shows that if Market Readiness is high but there is no technology, the risk is that the market need is not satisfied but continues to serve as an incentive for continued technology development to reach Technology Readiness. On the other hand, if the technology of a product is high, but there is no market, then the technology has the risk of being abandoned and losing all the investment in its development. Thus, although it is possible to develop successful innovative product through Technology Push, it is generally much more effective to develop the innovative product through Market Pull, which starts with the Market Readiness assessment based on an in-depth understanding of the needs and wants of the Final Adopter.

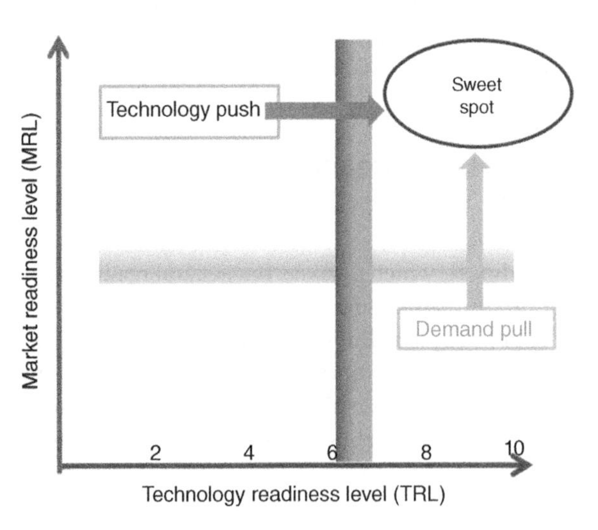

Figure 5.1 Conceptual Interactions Between Market Readiness and Technology Readiness [1]

Table 5.1 Early Warning of Market and Technology Risks

	TRL 1–5	TRL 6–10
MRL 6–10	Market available, no tachnology=> Market risk	Market and technology coherently ready
MRL 1–5	Market and technology in coherence, but not ready	Technology available, no market Technology risk

5.1.1 Marketing Test Bed (MTB): A Tool to Establish Market Entry and Track the Subsequent Interactions Between Market Readiness and Technology Readiness of an Innovative Product

The goal of an innovative product is to reach the "sweet spot" or "Window of Opportunity" for profit where Technology Readiness meets Market Readiness. However, to reach the goal, the product needs to start with market entry by understanding its adaptability to the Final Adopter. MTB is a useful tool to test the adoptability by asking Innovation Team and Final Adopter respective sets of questions based on the well-established Technology Acceptance Model (TAM). These questions focus on the two groups of technology acceptance criteria: Perceived Usefulness (PU) and Perceived Ease of Use (PEoU). Depending on the application domain, PU as well as PEoU might differ from the specificity of the addressed market segments and the segment-specific perception of usefulness and ease of use (EoU). Hence, dynamics of problem-solving capacity of an innovative technology causes impact on market segments' behavioral dynamics (e.g., Moore's Law in information technology). In addition to establishing marketability criteria for the product for market entry, MTB is also an important part of Customer Readiness, a key element of Market Readiness on the adoptability of subsequent technology developments of the product, to be discussed later in this chapter. As a result, MTB can also track the path of the product from market entry to the "sweet spot." The typical MTB questions include the following:

For the Innovation Team	
Innovativeness	Creativity & uniqueness?
Testability	Access to test data?
Controllability	Degree of autonomy?
Compatibility	Compatible to Adopter requirements?

Implementability	Implementable to Adopter ecosystem?
Assimilability	Degree of fit to Adopter?
For the Final Adopter	
Usage	Degree of match with Innovation?
Value	Perceived Economic-Ecologic-Equity values?
Risk	Perceived risk of investment in adoption?
Image	Perceived risk of image loss?
Routine	Perceived degree of routine or ease of use?
Assimilability	Degree of fit to supplier?

These questions serve to develop a transparent understanding of the required features of the innovative product for the application goal of the Adopter. Thus, MTB provides the following distinct advantages to understanding the practical interactions between Market Readiness and Technology Readiness for the actual adoption of an innovative product.

- MTB determines under which circumstances the Innovation Team's capabilities meet Adopter's needs.
- MTB serves as a knowledge translator between Innovation Team's language and Adopter's language.
- Resistance to adoption of innovation decreases as parties increase mutual understanding and generate mutual trust.
- MTB serves as a method to determine fulfillment of marketability criteria through reliable information to transform innovation from a creative idea to practical market entry.
- There are many tradeoffs among innovative attributes, such as creativity, testability, and understandability of the product which would cause Adopter's decision for rejection or acceptance with adoption and assimilation. MTB is an effective tool to clarify complementary and substitutional relations of innovative attributes and quality of related services such as training, maintenance, readability of instructions, spare part availability, etc.

5.2 Major Considerations of Marketing to External Adopter

Different from internal mutual marketing between Intrapreneur and Internal Supporter in an organization, marketing to External Adopter requires knowledge and considerations of the following:

5.2.1 Evolving Marketing Targets

Since External Adopter is outside the organization of the Intrapreneur, effective marketing requires the correct identification of the marketing target who makes decisions on the adoption and use of the innovative product. There are two major types of marketing targets:

- An individual marketing target or decision-maker: An individual decision-maker, such as a consumer, or a single decision-maker for an organization, like the CEO of a company in the role of Adopter.
- A group of marketing targets or decision-makers in the form of a committee with individual roles of influencer and power broker, such as a buying group or buying center, as a functional team of a business customer in B2B marketing.

Both types of marketing targets have greatly evolved due to increasing economic and intellectual resources in most parts of the world, which have resulted in rising individualism, diversifying needs and wants, and expanding ecologic and equity values among customers and employees alike. Understanding these changes will be the first step to attracting and engaging the External Adopter.

5.2.2 Advancing Marketing Tools

With the exponential expansion of information technology and behavior science, marketing tools have been rapidly advancing beyond the traditional tools, like word of mouth, individual relationship building, personal selling, and various print and video media advertisements, to include:

- *Digital advertising*: With the growth of the internet, digital advertising has become a major part of marketing. This includes display, search, and social media ads.
- *Social media marketing*: Social media has become a popular platform for businesses to promote their products. This includes creating content for social media platforms, running social media ad campaigns, and engaging with customers on social media.
- *Data-driven marketing*: With the ability to collect and analyze large amounts of data, marketers are able to create more targeted and personalized campaigns. This includes using data to understand customer behavior, segment customers, and create personalized content.
- *Marketing automation*: Marketing automation allows businesses to automate repetitive tasks such as email campaigns, lead nurturing, and social media scheduling. This helps businesses to streamline their marketing efforts and save time.
- *Mobile marketing*: With the growth of mobile devices, mobile marketing has become important. This includes mobile-optimized websites, mobile apps, and short messaging service (SMS) marketing.

- *Virtual and augmented reality*: Virtual and augmented reality technologies are being used to create immersive experiences for customers. This includes virtual showrooms, augmented reality product demos, and virtual reality training.
- *Experimental visualization of living lab demonstration* [3] showing PU and PEoU of innovative technology application. As a result, marketers can increasingly and precisely pinpoint the specific inner desires and hidden biases of the marketing target based on individual human needs, wants, and behavioral characteristics.

In the meantime, the well-established TAM [4] remains an important marketing tool for assessing the PU and PEoU of a technology. Together with TAM are the following important concepts of market acceptance [5]:

- Absorption Capacity, which assesses the effectiveness of the product marketing information to match the capacity of the customer for absorbing the information and represents the effectiveness of marketing communications [6]. To absorb the information means the ability of the customer to understand the usefulness and the usability of the innovation in the customer's productive ecosystem and the ability to use the innovation to attain customer's integrated Economic–Ecologic–Equity values [7, 8].
- Assimilation Gap, which is a key performance indicator (KPI) that measures the accumulated difference between number of acquisitions and number of implementations of the innovation [9]. There is a causality link between exaggerated willingness to buy based on euphoria and the reluctant willingness to implement and/or use caused by poor integrability into Adopter's ecosystem. This gap may be caused by Adopter's insufficient absorptive capacity to understand the innovation in all its necessary aspects and consequences, including integrability of innovation in existing Adopter's production ecosystem.

Assimilation gap is a major barrier to effective innovation adoption. Hence, the actual EoU is an important indicator of how and when a fully functional integration of the innovative product can be achieved for the Adopter's ecosystem. The full integration also includes the necessary behavioral integration of employees affected by the innovation, as an easy adaptation of working behavior to the innovation requirements is judged and perceived as high EoU. A high PEoU is also an integral part of PU of the innovation and a major element in convincing, committing, and maintaining the Adopter for both an individual or group acceptance of the innovation product in the working environment. Finally, PEoU also includes the customer expectation for the Ease of understanding of the use of the innovative product/system and its set-up, training of operating staff and maintenance, spare part management, and interfacing with existing sub-systems in the customer techno-system.

5.2.3 Major Changes in Innovation Marketing

As a result of these evolving marketing targets and advancing marketing tools, there have been major changes in innovation marketing:

- *Innovative business models rely more on branding, connectivity, and digitization (e.g., Airbnb, Uber, Amazon)*: Branding is a phenomenon based on emotional awareness to selected target groups motivating them to feel, think, and decide to purchase and use the offered product or services. To communicate and convey arguments for emotional conviction the ability to perceive and solve the customer's and user's problems. Branding can be understood as a collectively directed and emotionally informing communication tool based on the existing connectivity in highly developed economies. Success of branding depends, among other facts, on the degree of matching between the branding features and the touched receiver in his latent or emerging demand situation. A high EoU and the undoubted reliability of the supplier with given connectivity and an easily usable digital user interface to inform, execute, invoice, confirm payment, and deliver the ordered goods shows the effectiveness and efficiency of the real-time integration of branding, connectivity, and digitization, also covering global ethnical and cultural peculiarities. Furthermore, real-time connectivity and content digitization trade money for information availability and time savings. Thus, branding, connectivity, and digitalization will be especially important for globally attracting and engaging Adopter.
- *Innovation marketing will be increasingly supported not only by R-AI (Rational-Artificial Intelligence) but also by E-AI (Emotional-Artificial Intelligence) [10]*: An example of E-AI is the man–machine interactions to support human caregivers for elder care by providing individual E-AI featuring face, voice, and motion clustering with relevant vital parameters and semantic interpretation of emotional signals and contextual dialogues [11]. These new tools will be increasingly important for engaging and convincing an Adopter by generation of mutual trust.
- *High global connectivity requires innovative marketing to have systematic redesign of supply chain strategy considering EEE values for vital independence from noncompliant suppliers as risky members of the supply chain* (e.g., pharmaceutical supply chains, dependent on low-cost, but long-distance suppliers). Given increasing individual awareness and global connectivity, it is important to develop an integrated EEE value framework for innovation evaluation.
- *With intensifying global competition and increasing speed of innovation entering the market, the challenge for innovation marketing is to identify opportunity windows and complementary niches to be addressed systematically with controlled risk and reasonable Time to Market.*

5.2.4 Special Importance for Marketing Ethics to Individual Adopters

For individual adopters, increasing resources has greatly manifested diversified psychological needs and wants. On the other hand, advances in marketing tools and concepts have greatly increased the temptation for marketers to conduct unethical marketing for short-term gains by taking advantage of the information and capability inequity about new innovations of the marketee. If the innovation marketer aspires to long-term mutually beneficial relationship to commit and maintain the marketee, it must adhere to the following well-recognized composite marketing ethics:

1) *Honesty and transparency*: Ethical marketers are honest and transparent in their communication with customers. They provide accurate and complete information about their products or services and do not make false or misleading claims.
2) *Fairness*: Ethical marketers treat all customers fairly and do not discriminate based on factors such as race, gender, age, or socioeconomic status.
3) *Respect for privacy*: Ethical marketers respect the privacy of their customers and do not use their personal information for any purpose other than what they have consented to.
4) *Responsibility*: Ethical marketers take responsibility for their actions and the impact of their marketing efforts on customers, society, and the environment.
5) *Social responsibility*: Ethical marketers consider the impact of their marketing efforts on society and the environment and strive to make a positive contribution to both.
6) *Customer focus*: Ethical marketers prioritize the needs and wants of their customers, and do not engage in practices that exploit or harm them.
7) *Authenticity*: Ethical marketers are authentic in their communication and marketing efforts, and do not engage in deceptive or manipulative tactics.
8) *Compliance with laws and regulations*: Ethical marketers comply with all relevant laws and regulations, and do not engage in illegal or unethical practices.
9) *Continuous improvement*: Ethical marketers are committed to continuous improvement and seek to improve their marketing efforts and practices over time.
10) *Ethical leadership*: Ethical marketers are committed to ethical leadership, and lead by example by demonstrating ethical behavior and promoting a culture of ethics within their organization.

5.3 Market Readiness

As defined in Chapter 4, readiness is a system state of being fully prepared to execute a planned transition to a preferred future system state within required time, space, and resource availability. The success of Market Readiness of an innovation product is measured by the impact, i.e., the magnitude and duration of profitable adoption of the product through successful marketing communication based on a selected business model in the integrated EEE value framework for the target market.

5.3.1 Key Elements of Market Readiness

As shown in Figure 5.2, Market Readiness may be decomposed into four key elements: Demand, Competitive Supply, Customer, and Product Readiness. Each element deals with important necessary and sufficient conditions to achieve the overall Market Readiness for Intrapreneurship Innovation Management, preparing for market entry. Specifically, Market Readiness is a dynamic time-varying integration of these four key elements:

5.3.1.1 Demand Readiness

The first key element of Market Readiness assessment is the Demand Readiness for the product. To understand the demand, it is necessary to assess not only the need for External Adopter, which may be hidden or latent, and require the use of nescience analysis, but also how the demand is dependent on the specific components and their contextual configuration of the innovative problem solution, as further discussed below [12].

There are three major parts of Demand Readiness formation for innovation products as shown in Figure 5.3.

Wishful thinking is guesswork or presumption of an unstated problem, and a lack of knowledge on how to solve a fuzzy, unstructured, or ill-defined problem, almost not encountered in the daily lives of potential customers. The fuzzy feeling of "something is needed" emerges as the starting point of the path from latent wishful thinking via emergence to manifestation of demand for innovative products in the market. The feeling "something is needed" may also be caused by a "painful experience" (worst point) up to a "delightful hope" (best point) on a scale of "wishful thinking" from "how to avoid?" to "how to enjoy?"

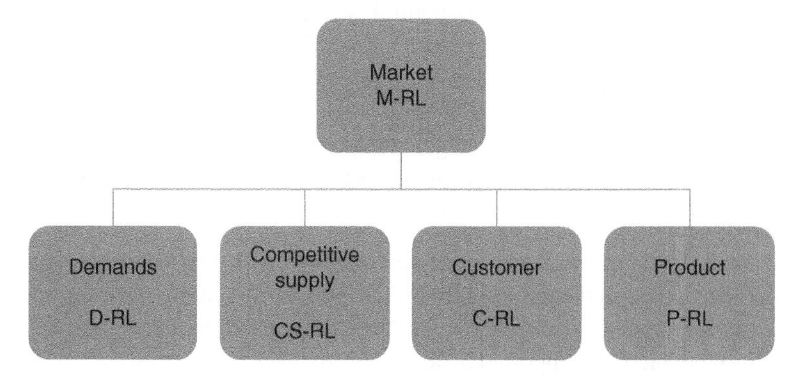

Figure 5.2 Key Elements of Market Readiness

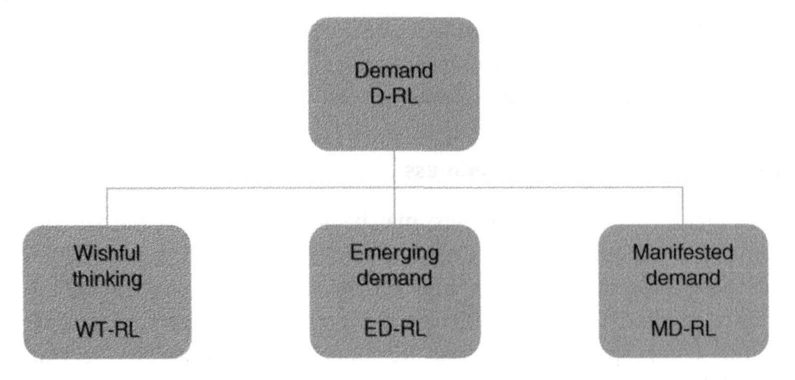

Figure 5.3 Major Parts of Demand Readiness

An example of wishful thinking as a hidden stage of demand is the search for solution to human structural bone disease.

The problem:

General: How to recognize early signals of human structural bone disease?

Specific: How to use X-ray pictures to significantly improve early detection of structural bone damage at initial stages of disease?

Discussion:

Based on the problem, there is a latent demand to increase accuracy and validity of bone diagnosis compatible with the existing X-ray image processing technology in medicine. Together with increased accuracy and validity, the operational diagnostic time and the degree of automation should also be increased. Furthermore, each pixel in the X-ray picture should contain the actual information on mineral bone density.

Solution requirements:

Finding an effective solution would require:

- Recognizing the initial stages of the structural bone changes caused by mineral bone density deficit ("how to avoid undetected mineral bone density deficit").
- Supporting accelerated development of pharmaceutical prototype products ("positive collateral usefulness effect").
- Preventing or at least retarding the malignant effects of low mineral bone density based on early warning signals (PU of early detectable mineral bone deficiency).

Wishful thinking solution or demand:

Instead of analyzing X-ray pictures in an analog mode with limited resolution power of the optical perception ability by the human eye of an experienced radiologist, the image analysis may be performed by pixel granularity resolution of the X-ray picture, supported by R-AI.

- *Emerging demand*

  At the second *stage* of *demand* formation, there is early anticipation that a solution is emerging for a perceived problem considered to be unsolvable until now. However, the emerging solution is still fuzzy and dependent on many different factors, such as:
  - societal and regional distribution of problem-induced knowledge.
  - frequency of communication with potential users and interested influencers and technology experts.
  - network connectivity effects like user groups and social network groups having a similar feeling that "something is needed" and "a solution is expected."

  The process of emerging demand is comparable with a percolation process of water trickling through porous material like sandstone. Fuzzy and weakly organized user groups might develop into future market segments by percolating early information on functional results and first experiences of the usefulness of an innovative solution. Early functional prototype of an innovation product would be tested and steadily improved through the feedback from Adopter's first experiences. This early feedback combined with the Intrapreneur's vision shapes the path from latent to emerging demand and can be understood as percolation process [13].

  In the above example of human structure bone disease detection, the emerging demand is the observation that each pixel of the X-ray shows a grayscale value which is a valid criterion for a specific mineral bone density value level. Hence the pixel-wise data analysis enriches the information content and delivers early-stage diagnostic facts, e.g., local distribution and patterns of critical gray scale values.

- *Manifested demand*

  At the last stage of demand formation, the versatility and adaptability of innovative problem solutions create a powerful space for continuing adaptation and improvement, and the demand for solution is now fully manifested for the prospective Adopter.

  Manifestation of demand may be seen as a growth process starting with first seeds of innovative ideas informally communicated and discussed in interest groups, followers, etc., following the communication of first results of functional tests showing the matching with latent demand for problem solution as well as expected and PU and PEoU, the two key technology acceptance criteria in the TAM.

  Example: Early warning information.

  For the bone structure disease detection example, algorithms of pattern recognition and R-AI-assisted diagnosis algorithms support the transformation of raw pixel data into a value-added early warning diagnosis system called Image Biopsy (IB) Lab, which has already been approved by the US Food & Drug

Agency [14, 15]. Thus, the stage of manifested demand is supported by the technology. The versatility of the IB Lab allows it to be applied to all different human bone types for diagnosis of different stages of potential structural bone disease. If human mineral bone density falls below threshold value, the IB Lab X-ray pictures will show risk of structural bone disease as a fully measurable, early warning hard data diagnostic information [16].

Demand Readiness is a lengthy process. From the starting point of wishful thinking and latent demand, through explored needs and fuzzy weak signals of emerging potential demand, until a set of early adopters form the milestones of Demand Readiness showing the path from early idea generation to professional market entry.

Example: Urban public cooling of street temperature.
As another example, the rising importance of EEE value framework, caused partly by the evolving individual and organizational needs for protection against global warming effects, has increased the Demand Readiness for innovative products providing countermeasures to these effects. A potential emerging demand would be for urban public activities to cool down street temperature with tree planting and cool water fog dispensers as implemented in many European cities in recent summers.

5.3.1.2 Competitive Supply Readiness

The next key element of Market Readiness is Competitive Supply Readiness, i.e., to what extent are competitors' products/marketable solutions available? The answer to this question requires competitor-centered market research and market observation. Accessibility to competitive data is not easy and often relies on informal information trading [17]. A KPI for this readiness is the Competitive Innovation Half-Life (C-IHL) that answers the question: "How long does it take to catch up to 50% of Innovation advantage held by the strongest competitor known to the innovator?" As shown in Figure 5.4, competitive advantages include advantages in quality, delivery terms, price, payment terms, intellectual property rights protection, ecological compliance, societal and cultural conformance, as well as cost, functional availability, and many other properties useful to the prospective customer. For example, during the recent COVID-19 epidemic, the supply chain advantages may have been more important than the innovation advantages due to shortage of skilled manpower and critical production components.

The concept of C-IHL assumes that the innovation lead is exposed to the speed of technological and scientific progress of the best-known (or maybe unknown) competitors. The addressed exposure results in a continuous decay of innovation lead. Each innovator is confronted with the risk of innovation lead decay. On the other hand, continuous innovation can also result in the growth of the innovation lead. Both decay and growth of the innovation lead may be represented by different functional forms, such as linear, exponential, elliptic, parabolic, hyperbolic, or logistic function, or a combination of these functions [18].

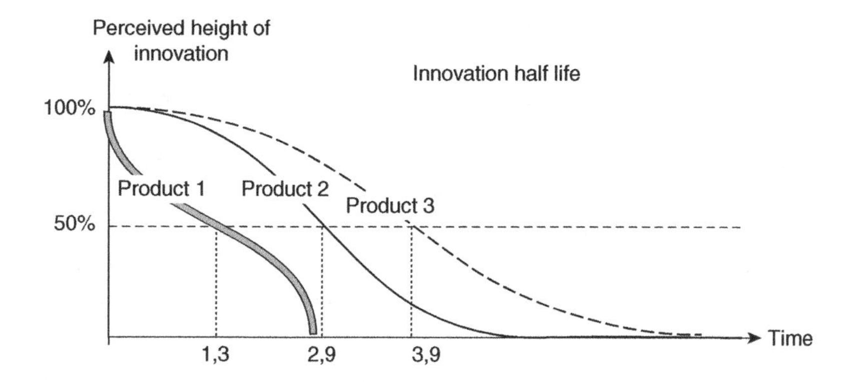

Figure 5.4 Illustration of Competitive Innovation Half-Life (C-IHL). C-IHL: How Long Does It Take Before 50% of the Innovation Lead is Taken up by the Best Competitor Known to the Innovator? Examples: Product 1: Short at 1.3 Years; Product 2: Average at 2.9 Years; Product 3: Long at 3.9 Years

Disruptive technologies may cause rapid decay of C-IHL for traditional technologies [19]. A striking example of an emerging class of disruptive technology is the family of quantum deep technology. Driven by the promising reduction of computing complexity with quantum computers and quantum algorithms for optimization of sensor positioning in cars, the new technology has rapidly sped up the replacement of the current optimization technology [20].

5.3.1.3 Customer Readiness

The next key element of Market Readiness is Customer Readiness, i.e., "Is the customer ready to adopt and use the product and its associated system (Training, Maintenance, Spare Part Management, Functional Modules)?" Here, we need to use the TAM to study both the absorption capacity and assimilation gap of the customer. Specifically, assessment of the absorption capacity requires the mapping of the innovation attributes and their quantitative or qualitative values into the customer's preference and goals, resulting in a preference judgment for the overall evaluation based on the integrated value framework:

- Economic evaluation shows how the innovation contributes to the economic goals in terms of
  - Sales quantity, contribution margin, return on sales, and market share.
  - *Productivity*: Output divided by input subject to commensurability of numerator and denominator.
  - *Reliability/Quality*: The longer the mean time between failure (MTBF) and the shorter mean time to repair (MTTR), the higher the service quality level.

- Ecologic evaluation shows how the innovation contributes to the attainment of ecological goals such as:
  - To avoid harmful gaseous output and required filter processes and their follow-up cost per unit.
  - To minimize non-recyclable and toxic waste in weight and percentage of output and follow-up cost of waste management.
- Equity evaluation shows how the innovation meets the societal norms and legally agreed standards of the affected groups in the society. KPIs for societal evaluation are required based on extensive observation and study of medical data for different age and education groups.

Moreover, based on TAM, there is a set of marketability criteria for Customer Readiness as shown in Figure 5.5. Figure 5.5 shows the relationship between MTB and acceptance or rejection of the innovative technology:

These marketability criteria are indicators of the innovation advantage. However, it is important to distinguish between announced innovation advantage by the supplier and perceived innovation advantage by Adopter. The difference between announced and perceived counts either for disappointment or euphoria.

WTP and WTU is heavily influenced by Adopter's personal acceptance of the innovation advantage. A sustaining innovation advantage based on not only technologic but also EEE values with high availability, reliability, and quality is of key importance to the Early Adopter and its ecosystem. The absorptive capacity of the target customer to learn, understand, and continually accept and receive advantages for the PU and the PEoU of the innovative product ensures a reinforcing effect for attaining the EEE values by the innovation supplying organization. Moreover, the innovator must communicate with the Early Adopter in language emphasizing the usefulness of the innovation and its competitive, even disruptive advantage. Specifically, innovators must provide customized messages with easy-to-understand examples and stories to motivate and support willingness to pay and to use decisions of the Adopter.

5.3.1.4 Product Readiness

The last key element of Market Readiness is Product Readiness, i.e., is the product ready for widespread use, which requires a proven compliance to EEE

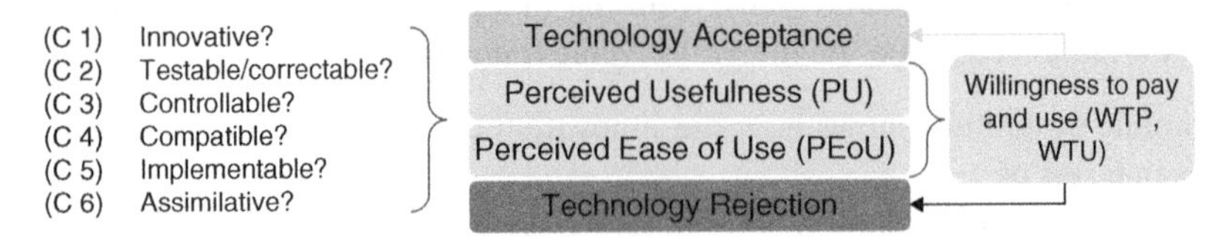

Figure 5.5 MTB Marketability Criteria (C1–C6). WTP: Willingness to Pay; WTU: Willingness to Use

standards? Product Readiness for widespread use requires safety and security against any adverse impact of material, malfunction, and influence on the user and its private and/or professional environment. Product Readiness for use/consumption thus shows a high degree of fulfillment of expected usefulness and expected EoU and compliance with the applicable regional or even global product standards.

5.4 Development of Market Readiness Indicators

Based on the above discussions, a set of questionnaires and checklists will be developed for innovation participants to assess through a set of indicators of the Market Readiness of an innovation product. These indicators will integrate the results from the following:

- Understanding the hidden and wishful, emergent, and manifested needs of the marketee *through deep immersion with marketee. Simulation of scenarios supported by Augmented Reality might unveil deeply hidden almost unconscious wishes.*
- Studying the absorptive capacity and assimilation gap of the marketee *through connectivity and digitization in real-time markets with AI-assisted analysis* [21].
- Assessing the innovation *complexity of market legal and societal conditions through* the *extended business model* based on the Triple Bottom Line and Triple Top Line (TBL/TTL) of the integrated EEE value framework expressed by the UN Sustainable Development Goals (SDG) [22].

Details of this development will be presented in Chapter 7.

Summary

This chapter has applied the unified approach to innovation marketing to the External Adopter. It has studied evolving marketing targets and advancing marketing tools including the absorption capacity and assimilation gap concepts for the application of the TAM. It then discusses the special importance of marketing ethics for the individual Adopter. The chapter next presents the key elements of Market Readiness: Demand, Competitive Supply, Customer, and Product Readiness. It concludes with an outline of the integrated studies for developing the indicators for Market Readiness as the inputs to the IRN© software.

Glossary

Absorption capacity: The behavioral and intellectual ability of the addressed customer organization to comprehensively understand, internalize, and put into action the communicated and taught innovation by the entitled members of the innovating organization.

Assimilation gap: The cumulative difference between acquisition and implementation of a purchased or acquired innovation in a specified period of time for a specified target group. (see [9])

Competitive Innovation Half-Life (C-IHL): The time that takes to catch up to 50% of Innovation advantage held by the strongest competitor known to the innovator.

Competitive Supply Readiness: The competitive supply is ready if the Competitive Innovation Half-Life is less than the required development time for a competing innovation.

Customer Readiness: Customer is ready if the assessment indicates that the individual Adopter in B2C market or the business decision-maker in B2B market would perceive sufficient EEE values and has sufficient absorption capacity to overcome any assimilation gap for paying and using the innovative product.

Demand Readiness: A demand is ready when potential customers (buyers, users, lessees) are willing to articulate their demand to the prospective suppliers and start to check the innovative competitive supply and come to a decision: yes or no to buy/to rent/to use, and to pay.

Emerging demand: The second stage of demand formation when there is an early anticipation that a solution is emerging for a perceived problem that is considered to be unsolvable until now.

Manifested demand: The last stage of demand formation when the versatility and adaptability of innovative problem solutions create a powerful space for continuing adaptation and improvement, and the demand for solution is now fully manifested for the prospective Adopter.

Marketability criteria: A set of tests for assessing the adaptability of a product by the intended market.

Market Readiness: A market is ready for an innovative product when an individual or business as buyer at the demand side of the market will screen the supply offer for the seller of the product and decide to fulfill a commercial transaction (buying, renting, leasing) at an agreed equivalent value (price, fee, etc.) to benefit from the product through the buyer's willingness to pay and to use.

Marketing Test Board (MTB): A tool for establishing the marketability criteria for market entry and testing the perceived usefulness (PU) and perceived ease of use (PEoU) of a product for market adoption.

Product Readiness: Product Readiness for widespread use is attained when the use/consumption of an innovative product meets the economic-ecologic-equity (EEE) value standards, shows a high degree of fulfillment of perceived expected usefulness and ease of use, and complies with all the safety, health, and security regulations against any adverse impact of material, malfunction, and influence to the customer and its private or professional environment.

Sweet spot or Window of Opportunity for profit: A condition at which both Market- Readiness and Technology-Readiness of a product are high and thus the ideal opportunity for sustainable profit.

Technology Acceptance Model (TAM): A model for evaluating the PU and PEoU of a product/service/technology.

Wishful thinking: At the beginning of demand formation, there is guesswork or presumption of an unstated problem, and a lack of knowledge on how to solve a fuzzy, unstructured, or ill-defined problem, almost not encountered in the daily lives of potential customers.

Discussions

Describe how marketing targets have changed over time.

- Describe how marketing tools have advanced over time.
- Describe how innovation marketing has changed over time.
- Discuss in-depth about marketing ethics: importance, guidelines, cultivation, and enforcement.
- Provide concrete examples of absorption capacity.
- Provide concrete examples of assimilation gap.
- Critique demand formation and readiness assessment.
- Further discuss Competitive Supply Readiness assessment and provide concrete examples on estimating C-IHL of a product.
- Further discuss Customer Readiness assessment.
- Further discuss Product Readiness assessment and any difference between goods and services.

References

1 Hasenauer, R., Weber, C., Filo, P., and Orgonas, J. (2015). Managing technology push through marketing testbeds: the case of the hi-tech center in Vienna, Austria. *Proceedings of PICMET Conference 2015* (2–6 August 2015), Portland, OR, USA. IEEE Catalog Number: CFP15766-USB, PICMET ISBN USB: 978-1-890843-32-8, pp. 99–127.

2 Day, G. and Freeman, J. (1990). Burnout or fadeout: the risks of early entry into high technology markets. In: *Strategic Management in High Technology Firms* (ed. M. Lawless and L. Gomez Mejia). JAI Press Inc.
3 Hasenauer, R., Ehrenmüller, I., and Belviso, C. (2022). Living labs in social service institutions: an effective method to improve the ethical, reliable use of digital assistive robots to support social services. *Proceedings of PICMET Conference 2022* (7–11 August 2022), pp. 1516–1524.
4 Davis, F. (1989). Perceived usefulness, perceived ease of use, and user acceptance of information technology. *MIS Quarterly* 13 (3): 319–340.
5 Venkatesh, V. and Davis, F. (2000). A theoretical extension of the technology acceptance model: four longitudinal field studies. *Management Science* 46 (2): 186–204.
6 Cohen, W., Levinthal, D., Absorptive capacity: a new perspective on learning and innovation, *Administrative Science Quarterly*, Volume 35, Issue 1 pg. 128-152, 1990.
7 Zahra, S. and George, G. (2002). Absorptive capacity: a review, reconceptualization, and extension. *Academy of Management Review* 27 (2): 185–203.
8 Zou, T., Ertug, G., and George, G. (2018). The capacity to innovate: a meta-analysis of absorptive capacity. *Innovation: Organization & Management* 20: 87–121. Published online: 06 February 2018.
9 Fichman, R. and Kemmerer, C. (2001). The illusory diffusion of innovation: an examination of assimilation gaps. *Information Systems Research* 10 (3): https://doi.org/10.1287/isre.10.3.255.
10 Czerwinski, M., Hernandez, J., and McDuff, D. (2021). Building an AI that feels: AI systems with emotional intelligence could learn faster and be more helpful. *IEEE Spectrum* 58 (5): 32–38.
11 Abdollahi, H., Mahoor, M., Zandie, R. et al. (2022). Artificial emotional intelligence in socially assistive robots for older adults: a pilot study. arXiv: 2201.11167v1 [cs.HC] (26 January 2022).
12 Weber, C., Hasenauer, R., and Mayande, N. (2018). Toward a pragmatic theory for managing nescience. *International Journal of Innovation and Technology Management* 15 (5): 1850045.
13 Goldenberg, J., Libai, B., Solomon, S. et al. (2000). Marketing percolation. *Physica A* 284: 335–347.
14 US Food and Drug Administration 510(k) Premarket Notification: K192109 KOALA IB Lab GmbH, Decision Date 11/05/2019.
15 Ljuhar, R. (2019). *ImageBiopsy Labs Receives FDA Approval for IB Lab LAMA*. G-MedTech News Center.
16 Ljuhar, R., Schön, C., and Ljuhar, D. (2014). Bone Assessment Redefined. In: *Vom Innovationsimpuls zum Markteintritt, Theorie, Praxis, Methoden* (ed. R. Hasenauer and W. Schildorfer), 152–160. Wien: WU-Universitätsverlag/Facultas. AG ISBN 978-3-7089-1255-4.

17 Schrader, S. (1995). Gaining advantage by 'leaking' information: Informal information trading. *European Management Journal* 13 (2): 156–163.

18 Hasenauer, R., Bednar, L.: Ein Verfahren zur adaptiven Prognose bei auslaufender Nachfrage (DECAY-MODELL) in: Hasenauer, R. (edt.), Modelle der computergestützten Marketingplanung, Verlag Anton Hain Meisenheim am Glan 1977. ISBN 3-445-01475-2, pp. 453–469.

19 Swayne, M. (2022). s. *The Quantum Insider.*

20 Pramanik, S., Vaidya, V., Malviya, G. et al. (2022). Optimization of sensor-placement on vehicles using quantum-classical hybrid methods. TATA 2206.14546.pdf. arxiv.org (29 June 2022).

21 Eljasik-Swoboda, T., Rathgeber, C., and Hasenauer, R. (2021). Automatic estimation of technology readiness and market readiness by the readiness navigator AI. In: *Innovation durch Natural Language Processing – Mit Künstlicher Intelligenz die Wettbewerbsfähigkeit verbessern* (ed. W. Bauer and J. Warschat), 253–268. Carl Hanser Verlag. ISBN 978-3-446-46262-5; E-Book-ISBN 978-3-446-46606-7; ePub-ISBN 978-3-446-46935-8.

22 Viscusi, W. and Aldy, J. (2003). The value of a statistical life: a critical review of market estimates throughout the world. *Journal of Risk and Uncertainty* 27 (1): 5–76.

6

Innovative Product Development: Creative Problem-Solving and Technology Readiness Assessment

6.1 Introduction

Based on the simple definition of innovation being an idea implemented with significant impact, creative idea generation to solve the problem arising from the need of the adopter is the starting point of the innovation process. On the other hand, to achieve significant impact, internal innovation in an organization must also solve problems arising from meeting the needs of the internal innovator and supporter, i.e., the Intrapreneur, and the Internal Supporter, for them to be willing to invest their resources in innovation implementation. This point cannot be overemphasized because, without creative problem-solving in response to the joint needs of all innovation participants, an idea is merely an invention that satisfies no more than the limited stimulation needs for intellectual/creative pursuits of the idea generator.

Understanding the needs of all participants is the prerequisite to problem definition and creative solution generation in innovation development. Understanding of the needs of Intrapreneurs and Internal Supporter has been extensively discussed in Chapter 4 in terms of Organization Readiness for internal innovation, as well as the needs of Final Adopter in Chapter 5 in terms of Market Readiness for the innovative product. This chapter will focus on the creative problem-solving and Technology Readiness for the development of an innovative product, i.e., whether the product provides an effective solution to the problems resulting from the needs and wants of innovation participants in general and Final Adopter in particular.

Creative problem-solving is not only the basis for innovation development but also an innate human ability as clearly observable from behavior of young children and the development of human civilization. However, research [1, 2] has

Intrapreneurship Management: Concepts, Methods, and Software for Managing Technological Innovation in Organizations, First Edition. Rainer Hasenauer and Oliver Yu.

Published 2024 by John Wiley & Sons, Inc.

indicated that there is a strong tendency for the ability of creative problem-solving to decline as humans mature. This chapter will first examine major barriers to the continued development of creativity in adults, then demonstrate creative problem-solving as a dynamic and trainable skill and present a systematic synthesis of the major approaches and tools to overcome these barriers, including a special discussion on creative problem-solving by a team. These discussions provide the conceptual framework and key elements of Technology Readiness for the development and implementation of an innovative product, which will be the basis for a set of assessment checklists to be used together with those of Organization and Market Readiness as inputs to the integrated Intrapreneurship Readiness Navigator (IRN©) software.

6.2 Main Barriers to Continued Development of Creativity

Extensive research and experience of the coauthors of the book indicate the following main barriers to continued development of human creativity into adulthood.

6.2.1 Social Control

In most societies, cultures, and organizations, the authorities in charge generally exert social control to sustain power, maintain stability, or provide protection to the group. As a result, obedience to authority and conformity to the expected group norms are generally encouraged or even enforced, starting in early childhood, through various forms of indoctrination, including cultural traditions, religious beliefs, and various types of formal and informal education. New ideas and independent creative thinking are often discouraged, forbidden, and even punished. As a result, creative problem-solving skill diminishes as individuals mature.

Moreover, in many organizations, creativity is discouraged by management because of the inherent aversion to failure, a shortage of resources, or a hierarchical organizational structure. Additionally, for highly regulated monopolies, such as public utilities or government agencies, new ways of problem-solving may even be explicitly prohibited due to restrictive organizational charters.

6.2.2 Risk Aversion

Prominent research studies [3, 4] have demonstrated the prevalent behavior asymmetry among adults in that they attribute a greater value to avoiding the pain of failure than to gaining the pleasure of success from trying something new. Coupled with prevalent social control, people generally become increasingly risk averse in creative problem-solving as they mature to avoid the perceived pain of potential failures, such as public humiliation, financial loss, and other negative consequences.

6.2.3 Cognitive Rigidity

Studies of the human thinking process [5–8] have indicated that a person's brain forms and modifies mental models of the perceived external world as well as the perceived impacts on the person's needs and wants from the interactions with the perceived external world. These perceptions are inherently imperfect and heavily influenced by genetic and environmental factors, while the modifications follow a crude, imprecise Bayesian logic. The modification becomes particularly difficult when the underlying realities are nonlinear and complex with significant uncertainties. As an example, a coauthor of the book has asked more than 6,000 college students over 23 years to intuitively estimate the probability that a person may be actually infected by a contagious disease with a 1% infection rate, after receiving a positive result from a test that has a 90% true positive probability (i.e., the probability that a person's test result is positive, given the person actually has been infected), and a 5% false positive probability (i.e., the probability that a person's test result is positive given the person actually has not been infected). Throughout the 23 years, over 80% of the students had consistently estimated the probability to be larger than 50%, while the mathematically correct probability based on Bayes Theorem is the ratio between the probability that the person has both a positive test result and the infection and the probability that the person has a positive test result, or $0.9(0.01)/[0.9(0.01)+0.05(0.99)]$, which is only 15%.

On the other hand, due to the evolutionary advantages of neurological energy conservation for behavioral efficiency [9], a person would often develop a fixed thinking process shaped by serendipitous past successes of intuitive mental models and perceived interactions with the objective reality. These reinforced mental models and thinking processes will generally lead to cognitive rigidity, or "psychological inertia" [10], i.e., strong habits in mental modeling, thinking, and problem-solving, which could be incorrect or ineffective. Moreover, cognitive rigidity can also be deliberately imposed on individuals for social control by powerful authorities in society through psychological conditioning or "brainwashing" with formal or informal indoctrinations.

Finally, people with strong beliefs and rigid cognitive processes often form close emotional ties with their fixed mental models and thinking processes and would strongly oppose the perceived threats from others with different but more creative models and processes. Such emotional attachment is also the reason why it is often difficult for adults to change their mental habits until drastic consequences occur due to large discrepancies between expected results from the rigid models and processes and the objective reality. In some cases, people would even modify their perceptions of the reality to sustain their mental models and thinking processes.

6.2.4 Skill Deficiency

Even though creativity is an innate human ability, with all the barriers, continued development of creative problem-solving requires continued skill training and reinforcement. In the following sections, we will first demonstrate creative problem-solving as a trainable skill and then present major creative problem-solving approaches and tools as specific skill-building countermeasures to these main barriers to creative problem-solving. Furthermore, given its rapid advances, we will also identify major opportunities where generative AI (artificial intelligence) can be used to enhance creative problem-solving skills.

6.3 Creative Problem-Solving as a Trainable Skill

Major studies have demonstrated that creative problem-solving is a dynamic and trainable skill due to the following facts:

6.3.1 Neuroplasticity and the Brain

Neuroscientific studies demonstrate that the brain exhibits plasticity, allowing for structural and functional changes in response to experiences. Research has identified brain regions and networks associated with creative thinking, and evidence suggests that training and practice can lead to modifications in these areas. For example, a study [11] used magnetic resonance imaging (MRI) to show increased gray matter density in the prefrontal cortex of participants who underwent creativity training compared to a control group. This finding supports the idea that creativity training can induce neuroplastic changes in the brain and enhance creative thinking.

6.3.2 Skill Development Through Deliberate Practice

The acquisition of expertise follows the principles of deliberate practice, a process involving focused effort, feedback, and repetition. Creative skills can be developed through deliberate practice, leading to improved creative thinking and problem-solving abilities. The role of deliberate practice in creative skill acquisition through a study on violinists has been demonstrated. [12]. It was found that expert violinists engaged in significantly more deliberate practice hours compared to intermediate and novice players. This suggests that deliberate practice can enhance creative abilities in specific domains.

6.4 General Approaches and Tools for Creative Problem-Solving

Based on the previous discussions, we have identified a set of general approaches and tools for effective skill development for creative problem-solving by overcoming the main barriers. Again, it cannot be overemphasized that the representative approaches and tools presented in this, and the next three sections are all highly process-oriented and interactive, and expert-facilitated training and continuous practices are necessary for successful applications.

6.4.1 Minimize Social Control Through Innovative and Nurturing Culture

Since cultures strongly affect human behavior and a social control-based culture severely inhibits creativity, development of creative problem-solving skills must start with a truly innovative and nurturing culture that encourages, facilitates, and rewards creative thinking. We have discussed the importance as well as the key elements of an innovative culture to Organization Readiness for Intrapreneurship development in Chapter 4. They are also the foundation for developing creative problem-solving skills.

6.4.2 Eliminate Risk Aversion Through Supportive and Stimulating Processes

In addition to an innovative culture, there also need to be supportive and stimulating processes for creative problem-solving to not only eliminate risk aversion but also to generate excitement among innovation participants. A well-established and widely applied process is facilitated brainstorming, which was first formalized by Alex Osborn [13]. To ensure a risk-free environment for the participants, the fundamental principles of brainstorming are:

- No criticism of any ideas
- Go for quantity
- Welcome wild ideas to stimulate far-out new mental models
- Combine and improve ideas

Additionally, it is a critically important duty for the facilitator to make brainstorming an exciting and enjoyable experience for the participants. A proven tool to accomplish this duty is to simply ask participants to respond to a new idea with "yes, and (additional new ideas)" instead of "yes, but (critique of the idea)." By expanding instead of critiquing a new idea, participants are both supported and stimulated, and the creative problem-solving process becomes fun and exciting.

Another effective tool to eliminate risk aversion for creative problem-solving among participants is to treat an idea that does not actually solve the problem as a "learning experience" rather than a failure. Even just changing the popular slogan of "failing fast" to "learning fast" will lend a positive tone and psychological encouragement to creative problem-solving.

6.4.3 Reduce Cognitive Rigidity Through Diversified Perspectives

As discussed earlier, cognitive rigidity is a natural result of conservation of neurological energy in humans. However, there are several proven tools that can be used to reduce the effects of this barrier to creative problem-solving:

- First is to follow the "Go for Quantity" principle of brainstorming to ask participants to generate more and potentially different perspectives on the problem and its possible solutions.
- Another tool is to include participants with distinctively different backgrounds, expertise, and experience, thus providing diverse perceptions and perspectives on a problem and its solution.
- Still another is to apply "framing" to the problem, i.e., to examine the problem from different perspectives to stimulate creativity. This tool has been widely used in behavioral research and creativity studies [14–16]. A simple application of this tool is shown in Figure 6.1a, b, where the rotation of an object presents two different perspectives of a duck and a rabbit.

6.4.4 Build Creative Problem-Solving Skills

There are many approaches and tools for building creative problem-solving skills. We will first summarize the traditional approach here, and then present three

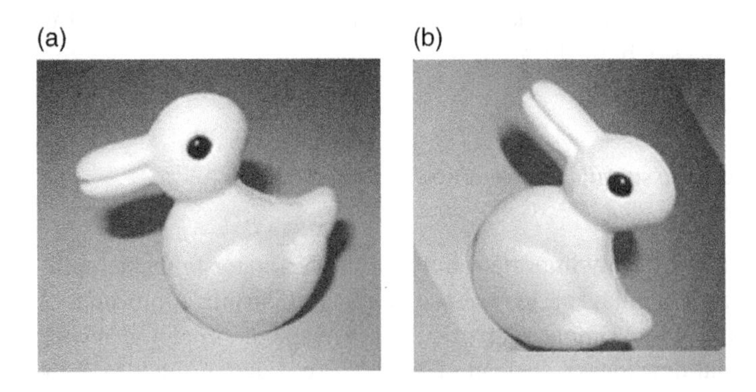

Figure 6.1 A visual example of framing. *Source:* With permission of D. Sheu

special approaches: Enneagram, TRIZ (Theory of Inventive Problem-Solving), and Nescience in separate sections.

6.4.5 Traditional Approach to Problem-Solving

The traditional approach is to apply the principles of logical thinking to creative problem-solving. In addition to the well-known induction and deduction, these principles include the following:

- *Abduction*: Abduction, also known as inference to the best explanation, involves the process of generating the most plausible explanation or hypothesis based on available evidence or observations. It is often used when faced with incomplete or ambiguous information.
- *Analogy*: Analogical reasoning involves drawing conclusions or making inferences based on similarities between different situations, objects, or concepts. By recognizing commonalities between different domains, analogical reasoning allows for the transfer of knowledge and understanding.
- *Causal reasoning*: Causal reasoning involves identifying and understanding cause-and-effect relationships. It enables individuals to infer causal connections based on observed patterns or evidence, allowing for predictions and explanations of phenomena.
- *Counterfactual reasoning*: Counterfactual reasoning involves contemplating what would have happened if certain conditions or events were different from reality. It enables individuals to explore hypothetical scenarios and assess the potential consequences of different actions or circumstances.
- *Inductive-statistical reasoning*: Inductive-statistical reasoning combines elements of induction and statistical analysis. It involves drawing conclusions based on generalizations from a sample or statistical data, incorporating probability and statistical reasoning to make predictions or judgments.
- *Dialectical reasoning*: Dialectical reasoning involves evaluating and resolving conflicting arguments or perspectives. It entails critically analyzing opposing viewpoints, identifying strengths and weaknesses, and seeking a synthesis or resolution that reconciles different positions.
- *Syllogistic reasoning*: Syllogistic reasoning is a form of deductive reasoning that involves applying logical rules to categorical statements or premises to derive a valid conclusion. It operates based on established rules of logic, such as the transitive property or the rules of syllogisms.
- *Hypothetico-deductive reasoning*: Hypothetico-deductive reasoning involves formulating and testing hypotheses to arrive at a valid conclusion. It follows a systematic process of proposing a hypothesis, making predictions based on the

hypothesis, and conducting experiments or gathering evidence to confirm or refute the predictions.

These additional principles complement induction and deduction and contribute to a broader understanding of logical thinking, enabling individuals to engage in more nuanced and sophisticated reasoning processes.

There is a wide range of specific applications of these principles to creative problem-solving. Many of them can be found in the cited references. We will provide a few examples below:

- Induction reasoning was used by Russian patent officer, Genrich Altshuller, and his followers to develop "Theory of Inventive Problem-Solving" [17] or TRIZ in Russian acronym, through a systematic and thorough study of about 40,000 technical patents to identify common patterns to form a set of fundamental principles, or pillars, or philosophies of creative problem-solving, which were later further expanded into comprehensive knowledge bases of successful applications of various problem-solving techniques. Through these principles and knowledge bases, successful solution techniques in many fields can be used to develop creative new applications in other fields. As a practical example, an inkjet printer developer needed to solve the problem of miniaturizing its printing technology, it was able to find a creative solution through a knowledge base of the integrated circuit (IC) industry that had successfully developed many miniaturization techniques [18].
- Biomimicry is a major extension of the analogy principle where creative problem-solving techniques, gained from observations of nature, are transferred to industrial applications. The following are two prominent examples:
  - The development of the widely applied fastening technology, Velcro, had its origin in the natural world, drawing inspiration from the burrs that cling to clothing and animal fur. In the early 1940s, Swiss engineer George de Mestral embarked on a hiking trip in the Alps and noticed how burrs effortlessly attached themselves to his pants and his dog's fur. This observation piqued his curiosity and set him on a path to replicate this mechanism. Following years of research and experimentation, de Mestral successfully recreated the hook-and-loop fastening system found in burrs. In 1955, he patented his invention and named it "Velcro," a portmanteau of the French words "velours" (velvet) and "crochet" (hook) and created one of the most successful applications of biomimicry [19].
  - As shown in Figure 6.2a, b, biomimicry of whale fins has provided creative solutions to improve the effectiveness of wind turbines. On the whale fin, large vortices are formed behind the troughs along the leading edge, whereas flow behind the tubercles forms straight streamlines. The effect of these flow patterns induced by the tubercles is to delay stall. It was observed [20],

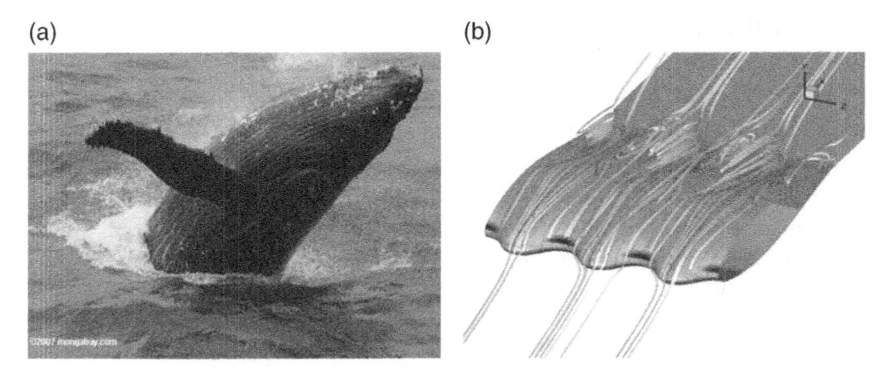

Figure 6.2 An example of biomimicry. *Source:* E. Pateron 2008/Conservation news/https://news.mongabay.com/2008/07/whale-biomimicry-inspires-better-wind-turbines (accessed 22 January 2024)

> "Engineers have previously tried to ensure steady flow patterns on rigid and simple lifting surfaces, such as wings. The lesson from biomimicry is that unsteady flow and complex shapes can increase lift, reduce drag, and delay "stall," a dramatic and abrupt loss of lift, beyond what existing engineered systems can accomplish."

In addition to these general approaches, we will introduce several somewhat lesser-known yet proven approaches for building creative problem-solving skills: a team-based creative problem-solving approach by the Enneagram, a systematic approach for problem-solving based on TRIZ, and the innovative concept of Nescience.

6.5 Team-Based Creative Problem-Solving Approach by the Enneagram (Based on contributions by Matt Schlegel)

As described in Chapter 4, the Enneagram system is a powerful tool to describe the behavioral dynamics and motivations of team building. The Enneagram system is also an effective problem-solving methodology, as the Enneagram dynamics are ordered in exactly the same way as the dynamics of human problem-solving. What distinguishes the Enneagram approach from others is that for each of the nine steps in problem-solving, there is a human personality dynamic perfectly tuned for that step. Team leaders can use the relationship between personality dynamics and problem-solving steps to predict in which steps of problem-solving the team will excel, in which steps they will struggle, in which steps they will get stuck, and which steps they will neglect altogether. Many team dysfunctions can be diagnosed based on this understanding, giving leaders a powerful tool to keep teams on track and moving smoothly through all problem-solving steps toward achieving the team's objectives. The Enneagram problem-solving approach can be used with any size team and can even be used by individuals.

6.5.1 Steps of the Enneagram Problem-Solving Approach

In the following descriptions of the problem-solving steps, we assume a typical work team of three to ten teammates, but there are no limits on team size.

Step 1: Problem and goal

While many problem-solvers will assert that the first step in problem-solving is clearly describing the problem, that is only half the picture. The other half of Step 1 describes the vision of the world in which the problem is eliminated. This vision serves as the goal of the problem-solving effort. Problem and goal are two sides of the same coin, and it is important to have your problem-solving team dedicate time to both aspects of this step. In practice, start with the problem description. Assure that each teammate has an opportunity to describe the problem from their unique perspective. Like the fable of the blind men in a room with an elephant, one may describe feeling crushed, one might complain about getting soaked, and another might complain about the smell. They may think they all have different problems when in fact they all have the same problem – the elephant in the room! It is important that teammates listen carefully to each other's perspectives during this session as listening creates a bond around a shared problem and serves to create an emotional connection among teammates by having a common enemy and a common cause to vanquish that enemy. When done well, Step 1 serves as a team building experience.

The other half of Step 1 is to create a shared vision of the future in which the problem is eliminated. Ideally, the vision-creation session is conducted the day after the problem description session. Again, each teammate shares his/her unique perspective of the future in which the problem is solved with all teammates listening carefully. The team creates a unified and detailed description of this vision which will serve as a beacon for the team as they move through the subsequent steps in problem-solving. Should new teammates join the effort in later steps, it serves the team well to revisit Step 1 with the new members, reviewing both the problem and goal, so that the new members quickly align their efforts with the team's objectives.

Step 2: Team commitment

Once the problem and goal are well understood, it becomes clearer to the team who will be involved in the effort, who will be required to work on the problem, who will be needed to support the team's efforts, and who will be impacted by the solutions. All related people become the effort's stakeholders. In Step 2, the team considers all stakeholders and how they must contribute to the effort. In a typical organization, these stakeholders may include leaders from different departments in the organization, operations, engineering, marketing, IT, finance, etc. There will also be executive sponsors (the Internal Supporters) who are responsible for providing the team with resources. There may be customers (the Final Adopter) of the effort who will be impacted. Ideally, the team identifies all stakeholders and records their roles and responsibilities in the effort, including those of the team members themselves. Again, if new members join the team in Step 2, it is worthwhile for the entire team to revisit Step 1 to ensure team alignment.

Step 3: Ideation

As soon as the mind is presented with a problem, it cares enough to want to solve it, it goes to work generating ideas for solutions. Step 3 in problem-solving is ideation. In this step, teams want to solicit the broadest possible set of ideas and perspectives from all stakeholders. Step 3 is not the time to judge or analyze ideas, it is simply the time to collect as many ideas as possible. The facilitator of this session may use a warm-up exercise to get the team's creative juices flowing. An example of a warm-up exercise would be to have the group think of all the similarities between a cat and a refrigerator. Once everyone has had a chance to contribute to the fun warm-up exercise, then turn the group to the problem at hand. Often a problem will be multifaceted. You can generate ideas around each facet of the idea. Remember, keep it fun and positive – doing so will enable your team to create the richest setup ideas.

Step 4: Emotional reaction

Every time you hear an idea, you have an instantaneous emotional reaction. That idea is great! Or, that idea stinks! We cannot help ourselves. And our reaction happens before we have even had a chance to reflect on the idea. This emotional reaction is Step 4 of problem-solving. Since Step 4 happens concurrently with Step 3, in practice they cannot be separated. Since you want to maintain a positive environment during the ideation session, the facilitator must coach the team that the emotional reactions – positive and negative – are perfectly natural and normal. Remind the group that during ideation the goal is to maintain a positive environment. If a teammate does have a negative reaction to another teammate's idea, then have them use that reaction to create another different idea that they like better. Doing so will allow the continuous generation of new ideas without creating contradiction and conflict. Once the ideation session is complete, generally no more than 90 minutes, then compile all the ideas and send them out to the group. Have each teammate rank the ideas in order of favorability. By democratically deciding the most favorable ideas, the group will naturally select the ideas with the most positive energy, the ideas they have the most emotional vesting in seeing come to fruition. Since there is still much work to do, selecting the most positive ideas will ensure the team has the positive emotional energy to see the problem-solving effort through to successful completion.

Step 5: Logical analysis

Once the team selects a small set of positive ideas to pursue, it is time to validate the efficacy of the ideas. In Step 5, the team conducts the logical analysis of each idea, including cost-benefit and pro–con analyses. After determining the best idea or ideas, the team may want to prototype candidate solutions to select the most promising candidate that is most likely to succeed in solving the problem and achieving the goal. The team may want to select the best two ideas or order to have one as a backup.

Step 6: Planning

With the most promising candidate in hand, the team moves to Step 6 and creates a detailed plan to implement the solution. The team considers all resources necessary to complete the project – budget, materials, expertise, tools, etc. Importantly, the project is mapped out on a timeline creating a project schedule highlighting when resources will be needed during the project. The team can also identify risks and contingencies. Here the team can also create a backup plan – Plan B – to be ready should the primary plan – Plan A – encounter an insurmountable barrier. Once the team develops the detailed plan, it has the information necessary to present to the stakeholders.

Step 7: Promotion

In Step 7, the team presents the plan to the stakeholders for review and approval. If stakeholders have been kept informed of progress to this point, this step could be merely a formality but still serves as a milestone to get approval to move forward with the project. The team can benefit from a presentation crafted using the order of the Enneagram problem-solving process, starting with Step 1, the problem statement and vision statement. The team can explain how the stakeholders were identified in Step 2 and show how they generated and selected the candidate ideas in Steps 3 and 4. They can demonstrate how they validated the ideas in Step 5 and built a plan around the best idea in Step 6. By walking stakeholders step-by-step through the process, the team brings the stakeholders to the point where they want to move forward and initiate action to solve the problem. Here the team gets approval to move to implementation.

Step 8: Implementation

Steps 1 through 7 have involved discussion and deliberation. Step 8 is the time for action. Once the team receives approval in Step 7, implementation can begin. In practice, the time the team will spend in Step 8 will vary based on the plan and the availability of the resources necessary to implement the solution. Remember, the plan formulated in Step 6 serves more as a guideline than a rule. Step 8 requires flexibility to navigate the various challenges that may arise during implementation. Should a big challenge arise that stalls the project, it can be helpful to use the Enneagram problem-solving process again to help solve the new problem. In Step 8, the team keeps driving forward until they reach their goal.

Step 9: Debriefing

In this last step, the team reviews the implemented solution with all the stakeholders. For complex implementations, some stakeholders may not be completely satisfied with the results. The team gathers feedback and assesses the efficacy of the implementation from each stakeholder's perspective. The team also compares actual results with the goals and the vision. Is the problem completely solved? Have new, unforeseen problems arisen? Step 9 is to ensure that stakeholders are sufficiently satisfied. If not, then record any new problems and determine if they are of sufficient consequence and magnitude that they, too, need to be solved. If they do, the team moves back into Step 1. And that is why the Enneagram is represented as a circle with Step 9 moving into Step 1 – the Enneagram represents a process of continuous improvement.

6.5.2 Personality and Problem-Solving

The Enneagram is at once a powerful personality dynamics system and a nine-step problem-solving methodology with each of the nine personality dynamics tuned for a specific step in problem-solving. Table 6.1 summarizes the relationship between personality and problem-solving. Briefly, Enneagram Type 1 focuses on how things should and should not be, intuitive identification of a problem, and vision for a world free of that problem. Enneagram Types 2, 3, and 4 are in the Heart or Emotional center of the Enneagram. Type 2 introduces the emotional aspect of caring to want to solve the problem. Type 3 suppresses emotions and generates ideas for successful outcomes. Type 4 responds emotionally to the environment and reacts to ideas, positively and negatively. Most common problem-solving methodologies neglect the emotional aspects of human nature; the Enneagram problem-solving approach centers human emotions in Steps 2, 3, and 4.

Table 6.1 Enneagram Problem-Solving Steps and Related Personality Dynamics

Enneagram Type	Personality Dynamic	Problem-Solving Step
1	Discerns how things should and should not be	Describe in detail the problem, and create a vision/goal statement
2	Connects emotionally to others by caring and helping	Identify commitment of all stakeholders
3	Seeks to be recognized for success	Generate a rich set of ideas for solutions
4	Responds emotionally to environment	Filter ideas by those viewed most positively by stakeholders
5	Collects and analyzes information	Conduct pro–con analysis and validate idea efficacy
6	Plans ahead to avoid danger	Create project plans including schedules, budgets, resources, and risk mitigation
7	Seeks to create a fun, positive community	Promote plan to stakeholders and get approval to solve the community's problem
8	Takes action to secure and control their environment	Implement the plan, solve the problem, and achieve the goal
9	Works to reduce conflict in environment	Review outcomes with stakeholders to ensure satisfactory implementation

Next comes the Head or Thinking Types 5, 6, and 7. Type 5 serves the analytical aspects of problem-solving. The risk averse Type 6 provides planning and risk mitigation. Type 7 connects with stakeholders and assures that everyone likes the plan and wants to move forward with implementation. Moving back to the Gut or Intuitive types, Type 8 initiates action and leads the implementation. Type 9 maintains harmony during and after implementation, ensuring that the implementation satisfies the needs of the stakeholders.

Each Enneagram type brings a specific strength and energy to team problem-solving and highly effective teams account for each and all of these dynamics.

6.5.3 Team Diagnostics and Special Strength for Problem-Solving

Team leaders who take a systematic approach to team problem-solving would ensure that no steps are missed, thereby increasing the likelihood of successful project outcomes. Failing to do so can lead to unpredictable outcomes and the need to revisit steps. Teams that fail to take a systematic approach subject the project to the whims of dominant personalities on team. People want to play to their strengths and will jump to the problem-solving step that resonates with them. Many team dysfunctions can be diagnosed by an over- or under-emphasis on a given step. For instance, *paralysis by analysis* arises when Type 5 dynamics overwhelm the effort and cause the team to get stuck in Step 5. *Ready, Fire, Aim* describes a team with an abundance of Type 8 dynamics wanting to skip steps and get straight to implementation in Step 8 without having thought through a plan. Having a completely balanced team is exceedingly rare and having too much or too little of the nine Enneagram dynamics is the norm. However, team leaders can overcome the influence of team imbalance by informing team members of the systematic approach the team will use and then sticking to that approach throughout the effort. While strong personalities will still try to jump forward (or backward) to their point of strength, the team leader can use the systematic problem-solving framework to bring them back to the step at hand.

In sum, as a problem-solving system, the Enneagram distinguishes itself with its intimate relationship to human nature – our strengths, weaknesses, emotions, intuitions, and thoughts. As such, it serves as an ideal platform for team-based problem-solving by providing a rich set of tools to keep teams on track to solving the collective problems and achieving the collective goals. More details of this approach can be found in *Teamwork 9.0. Successful Workgroup Problem-Solving Using Enneagram* [21].

6.6 Systematic Approach for Creative Problem-Solving as Represented by TRIZ (Based on contributions by D. Daniel Sheu)

As discussed in Section 6.4, TRIZ was developed by Russian patent officer, Genrich Altshuller, and his followers. In addition to providing a systematic structure for analyzing problems and their possible solutions, TRIZ emphasizes that a problem be first converted to the basic abstract form, or model of problem, so that it can be examined in many different perspectives to enable the idea generator to break out from the restrictions of existing mental models and fixed perspectives, or psychological inertia in TRIZ, and apply different tools to solve the problem. Moreover, although not specifically emphasized, TRIZ practitioners generally view identifying the need as the first step for solving the "correct" problem. This section presents a summary of the major philosophies of TRIZ with the latest developments [22].

6.6.1 A Predictable Pattern of Technology Advancement

After analyzing over 40,000 technical patents, Altshuller observed a predictable pattern of technology advancement through repeated cycles of birth, growth, maturity, decline of a technology, and its replacement by new technology. Such patterns can be used to identify through scenario analysis the next-generation technology developments.

6.6.2 Contradiction

In studying different approaches to problem-solving, Altshuller realized that contradiction is the fundamental barrier to all technical advancement, as many people succumb to the psychological inertia of seeking a compromising solution to the contradiction dilemma. By analyzing successful solutions to these contractions in innovative patents, he derived 40 inventive principles for solving engineering problems. These principles have been further expanded and extended to non-engineering fields such as business management. The contradiction philosophy serves two purposes: one is to use past successful experiences to suggest potential directions for searching the solutions to a problem, and the other is to enable an idea generator to identify contradictions that are unknown or nonobvious to others and thus the opportunities to innovate or resolve those contradictions by innovation.

6.6.3 Functionality–Value Hierarchy

Once a contradiction is identified, the idea generator needs to examine the Function–Value hierarchy to break out the narrow restrictive psychological inertia of focusing only on the most directly obvious problem at hand. Specifically, while most technical product developers focus on the product they produce, TRIZ's Functionality–Value concept focuses on the function of the product, not the product itself. Therefore, the mission of the innovator should be to achieve Functionality without the focus on the physical product. This brings a distinct paradigm shift to allow engineers to be free from the existing product constraints and explore many different products to achieve the same or better Functionality. This is level 1 "think outside the box" mentality.

Sheu [23] has further extended the Functionality concept to the value of the product which is the real purpose of the product. If one can achieve the same or higher level of value, the function of the current level can be eliminated or replaced with other function. Sheu further proposes this Value–Function–Effect–Component hierarchy. Solving a problem can be done at different levels of the hierarchy:

1) Achieving higher level value instead of the current one.
2) Achieving current value with different functions.

3) Achieving same function with different action principles (effect).
4) Achieving same action principle with different components.

The higher the level is satisfied, the higher level of creativity or innovativeness that can be achieved. This concept can also be used in systematic patent circumvention, enhancement, and regeneration.

Take washing machine as an example. The main function is to wash clothes, and the hierarchy for this example can be represented as follows: The function of "cleaning clothes" has the value of "clothes being clean". The following different levels of innovative designs can thus be considered:

1) *Achieving a different or higher level value* instead of "clothes being clean": If the value of "clothes being clean" is to please a new friend, we may instead prepare a small gift to please this friend. This satisfaction at a different or higher level value can provide a paradigm shift to new innovation opportunities for a different industry.
2) *Satisfy at the value level instead of functional level*: Instead of using a washer to perform the "cleaning clothes" function, we can develop an innovative dirt-repellent cloth to satisfy the value of "clothes being clean" (never get dirty). Such innovation is actually currently available through the ultra-ever-dry coating material from UltraTech International, which applies nanotechnology to coat a thin film on the surface of cloth to repel water, oil, other liquids, and particles.
3) *Satisfy the Functionality but use different action principles* to achieve the same Functionality. For example, instead of water physically dissolving dirt to clean clothes in the washing machine, we can use chemical reactions, ultrasound vibration, and many other innovative alternatives to clean clothes.
4) *Explore different components or systems to achieve the same action principle or function:* For example, instead of using water to carry detergent, we may use other liquids, such as alcohol or steam, to carry detergent and obtain the same results. Many people will focus on this level of problem-solving. However, this is at the lowest level of innovativeness. The "Value–Function–Effect–Component hierarchy" allows us to resolve problems systematically and comprehensively at various levels of innovativeness. On the other hand, resolving problems at the value level almost always creates a new technology or industry that may well eliminate the current technology or industry.

6.6.4 Ideality

Once a specific Functionality–Value level for the creative solution to a contradiction is identified, Altshuller introduced the philosophy of ideality to provide the ultimate goal for the solution.

Ideality derives from "the ideal machine," an ultimate system that has all its parts performing at the greatest possible capacity. Ideality is a measure of how close a system is to the best it can possibly be, that is, the ideal machine or the ideal final result (IFR). Ideality is defined as shown in equation below. The benefits are the useful functions provided by the system while costs and harms are its unwanted outputs or waste products, also regarded as harmful functions of the system.

$$\begin{aligned}\text{Ideality} &= (\text{Perceived Benefits}) / (\text{Costs} + \text{Harms}) \\ &= (\text{Useful functions}) / (\text{Harmful functions})\end{aligned}$$

The best value of ideality is infinity when the IFR is achieved with cost or harm (cost + harm = 0). Even though in reality the infinity level may not be achievable, it provides the direction the system designer needs to strive for. Ideality prompts idea generator to put the psychological inertia of pre-set mental constraints aside and think of ways to achieve the ideality as the goal. If the ultimate ideality cannot be achieved, a systematic step-back process can be used to find second choice or third choice, etc. The results are often better than those solutions developed by considering constraints first. Ideality also includes self-service principles, such as self-cleaning, self-cooling machines, etc.

6.6.5 Resources

For TRIZ, a resource can be any substance, field (energy), function, attribute, space, time, information, or even vacuum, void, or "nothing" that can be used toward some purpose. There are two key ideas under this concept:

- *Useless to Useful (U2U)*: Use resources that are not used, discarded, not designed to, or even seem irrelevant to problem-solving, but can be used to achieve desirable functions. For example, the traffic of Tokyo metro station has more than four million commuters every day, and the energy generated by passengers stepping on the floor was not utilized. The subway engineers installed piezoelectric materials on the floor, and they were able to convert the unused stepping energy into electricity to be used at ticket gates and other signage.
- *Harmful to Helpful (H2H)*: Convert harmful substances into useful resources. For example, thermal power generation has traditionally transformed "thermal energy into electrical energy" through the production of steam from fuel combustion to drive the turbine and rotate the generator. The process of burning coal or oil emits flue gas which contains particulates, sulfur dioxide, nitrogen oxides, and carbon dioxide, which will not only cause air pollution but also endanger human health. Meanwhile, these extremely high-temperature gases

or dusts will also form thermal pollution. In the past, people would build a tall chimney to let the particle pollution and thermal pollution reach a level of hundreds of meters above ground to dilute the pollution density so that when the particle pollutants and temperature pollution reach the ground level, they become bearable by humans. Now, a new technology takes the flue gas to be discharged to the atmosphere to an electrostatic dust collector so that more than 99.9% of the particles are collected and can be pressed into building bricks. As a result, harmful pollutants become useful building materials. The remaining high-temperature but cleaned flue gas is then fed into a heat exchanger to pre-heat the fresh air before it goes into the furnace to burn fuels. As the fresh air now carries more energy than before, less fuel will be needed to heat water into steam and produce more electricity. In summary, with the new technology, "harmful particulate pollution is converted into useful building materials," and "harmful thermal pollution is converted into more electrical energy."

The following are recent developments in TRIZ philosophies.

6.6.6 Alternative Space–Time–Domain–Interface Perspectives

The Space–Time–Domain–Interface (STDI) is an extension of Mann's Space–Time–Interface by Sheu [24] to cover more comprehensive situations. STDI implies that to solve a difficult problem, we often need to jump to different spaces, times, domains, and/or interfaces for different perspectives to understand the problem better and stimulate us to think of new solutions to the problem. The "framing" example in Figure 6.1a, b presented earlier illustrates the point: Figure 6.1a shows a duck, but when it is turned 90° clockwise it becomes a rabbit as shown in Figure 6.1b. These two pictures are exactly the same but are seen differently by turning 90°.

An example of the more sophisticated domain change is the usage of the famous Fourier transfer that converts a function in the time domain, $f(t)$, into a corresponding function in the frequency domain $g(w)$. A system's behaviors in the time domain represented by $f(t)$ may be modeled into a set of differentials and/or integral equations that are difficult to solve. Applying Fourier transfer, the original time domain equations become equations in frequency domain and the complex differential and integral equations are transformed into corresponding arithmetic equations with simple addition and subtraction operations.

The use of simulation can also be considered as using an alternative domain to transform and solve a problem in the primary domain. For example, installing a new factory layout usually requires large capital investment and long implementation time. If the physical layout of a design turns out to be undesirable, it would incur large financial and time losses, and undesired complexity increases. Changing the physical domain into the simulation domain of a virtual software model offers the benefits of low cost and simple modifications. This is the power of changing perspective to a different domain for problem-solving.

It is useful to note that a number of original TRIZ tools contain the essence of STDI. For example, the first step of the TRIZ problem-solving process is "converting Specific Problem into model of problem" which can be considered as seeing the problem not in its original form (specific problem) but in an abstracted form (model of problem), to allow different perspectives based on different STDI. Many other TRIZ tools such as contradiction matrix (seeing problem from parameters perspective), functional analysis (seeing problem from function–component relationship perspective), etc. may thus be viewed as extensions of the STDI concept.

6.6.7 System Transfer

When a problem occurs within a system, the issue of solving the problem can be transferred from within the system to an unrelated system to allow more innovative solutions. An example is the problem of needle threading in Figure 6.3. When threading, the eye of the needle is required to be large for easily putting the thread through the needle eye. However, the needle eye is also required to be small in order not to damage the clothes. As one of the system transfer solutions, we can meet these conflicting requirements by transferring the problem from the traditional needle to a new system as shown in Figure 6.3. In this new system, the hole of the threading device is large when the force is not applied, and smaller enough to get into the pinhole when the lateral compression force is applied.

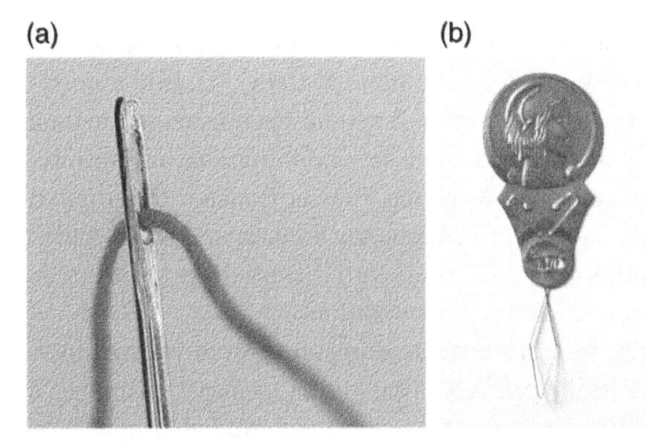

Figure 6.3 (a) Needle Threading. (b) Solution Based on a New System. *Source:* With permission of D. Sheu

6.7 Exploring Nescience for Creative Problem-Solving

One of the coauthors of this book and his colleagues have engaged in research on exploring the concept of Nescience for creative problem-solving [25]. The following is a summary of the findings.

Nescience refers to the state of lacking knowledge or awareness about a particular subject or problem. It is often considered an obstacle to effective problem-solving, as it implies a lack of information or understanding. However, emerging evidence suggests that Nescience can also be seen as a valuable resource that can lead to novel insights, creative solutions, and breakthrough innovations. Through a deliberate recognition of the lack of knowledge and an active pursuit of understanding, Nescience can spur creative thinking by challenging existing assumptions and prompting individuals to explore new perspectives. It can serve as a catalyst for innovative solutions by encouraging individuals to question established norms and think outside the box. Moreover, Nescience can facilitate exploratory problem-solving by fostering curiosity, experimentation, and the discovery of alternative approaches.

6.7.1 Nescience as a Catalyst for Creative Thinking

One key aspect of Nescience in problem-solving is its potential to stimulate creative thinking. Nescience disrupts established patterns of thought and challenges existing assumptions, encouraging individuals to explore new perspectives and generate innovative ideas. When confronted with a lack of knowledge or understanding, problem-solvers are more motivated to fill the gaps through creative thinking processes such as divergent thinking, analogical reasoning, and conceptual blending. They are more likely to generate original and unconventional solutions compared to those who rely solely on their existing knowledge. This highlights the importance of recognizing and leveraging Nescience as a catalyst for creative thinking.

6.7.2 Nescience and Exploratory Problem-Solving

Exploratory problem-solving involves navigating complex and uncertain problem spaces by exploring multiple potential solutions and learning from feedback. Nescience plays a crucial role in this process by promoting curiosity, experimentation, and the discovery of alternative approaches. When confronted with Nescience, individuals are more likely to engage in exploratory behaviors, such as conducting research, gathering diverse perspectives, and experimenting with different strategies. A study [26] explored the relationship between Nescience and cognitive flexibility, a cognitive process essential for exploratory problem-solving. The findings revealed that individuals with higher levels of Nescience demonstrated greater cognitive flexibility, enabling them to adapt their problem-solving strategies and generate more effective solutions. These results underscore the value of embracing Nescience to facilitate exploratory problem-solving.

6.7.3 Nescience and Problem Formulation

Effective problem formulation is crucial for successful problem-solving. Nescience can play a vital role in shaping problem formulation processes. When individuals are aware of their Nescience, they are more likely to engage in extensive problem exploration, breaking down complex problems into manageable subproblems and considering various perspectives. By embracing Nescience during problem formulation, problem-solvers can ensure comprehensive problem understanding and identify alternative problem representations that may lead to more effective solutions.

6.7.4 Challenges and Ethical Considerations

- *Cognitive biases and Nescience*: While Nescience can offer valuable opportunities for problem-solving, it is essential to be aware of cognitive biases that can hinder its effective utilization. Confirmation bias, for example, may lead individuals to seek information that confirms their preexisting beliefs rather than actively addressing Nescience. Overcoming cognitive biases requires a conscious effort to acknowledge and challenge one's assumptions, actively seek diverse perspectives, and critically evaluate new information.
- *Communicating Nescience effectively*. Effectively communicating Nescience is crucial for fostering collaboration, managing expectations, and encouraging further exploration. Communicating Nescience requires transparency, clarity, and an emphasis on the importance of ongoing learning and adaptation. It is essential to convey that Nescience is not a weakness but an opportunity for growth and innovation. Engaging stakeholders and facilitating open dialogues can help bridge the gap between Nescience and collective problem-solving.
- *Ethical implications of leveraging Nescience*: Leveraging Nescience in problem-solving processes raises ethical considerations. It is essential to consider potential risks, unintended consequences, and ethical dilemmas that may arise when embracing uncertainty and exploring uncharted territories. Responsible and ethical use of Nescience requires a balance between the pursuit of knowledge and the potential impact on individuals, communities, and the environment. Ethical guidelines, informed consent, and ongoing evaluation are crucial in navigating the ethical dimensions of leveraging Nescience.

In conclusion, Nescience, often considered an impediment to problem-solving, has the potential to be a valuable resource when harnessed effectively. By embracing Nescience, individuals can foster creative thinking, engage in exploratory problem-solving, and make adaptive decisions in uncertain environments. As we navigate increasingly complex and dynamic challenges, recognizing and embracing Nescience can open new pathways for innovation and generate novel solutions to address the problems of today and tomorrow.

6.7.5 Case Examples of Applying Nescience to Problem-Solving

- *NASA's Curiosity Rover Mission*: NASA's Curiosity Rover mission to Mars provides an excellent example of applying Nescience to problem-solving. When the mission was launched, scientists and engineers acknowledged the vast knowledge gap about the Red Planet and the challenges associated with exploring its surface. They recognized the Nescience regarding Mars' geological history, potential for supporting life, and the precise landing conditions.

  By *embracing* Nescience, the mission team designed the Curiosity Rover to be equipped with advanced instruments and sensors capable of collecting vast amounts of data. They employed a problem-solving approach that involved continuous exploration, experimentation, and data analysis. As the Rover roamed the Martian surface, it gathered valuable data about the planet's geological composition, atmospheric conditions, and potential for past habitability.

  The mission's success relied on embracing Nescience as a driving force for knowledge acquisition and problem-solving. By acknowledging what was unknown, the scientists and engineers were motivated to push the boundaries of scientific understanding, uncovering new insights, and making significant discoveries about Mars' history and potential for supporting life.
- *Google's Self-Driving Car Project*: Google's self-driving car project, now known as Waymo, is another compelling example of applying Nescience to problem-solving. When the project was initiated, the concept of autonomous vehicles presented numerous challenges and uncertainties. The project team recognized the Nescience surrounding self-driving technology, including regulatory and legal frameworks, safety concerns, and public acceptance.

  To address these challenges, the project team adopted a problem-solving approach that embraced Nescience as an opportunity for learning and innovation. They invested heavily in research and development, gathering extensive data on road conditions, traffic patterns, and driving behaviors. Through rigorous testing and analysis, they aimed to understand and overcome the Nescience associated with autonomous driving.

  By *acknowledging* the gaps in knowledge and actively pursuing solutions, Google's self-driving car project made significant progress. The team developed advanced machine learning algorithms, sensor technologies, and safety protocols to improve the capabilities and reliability of autonomous vehicles. Today, Waymo's self-driving cars have successfully logged millions of miles on public roads, showcasing the power of embracing Nescience in solving complex technological challenges.

- *Early warning detection of structural bone disease*: Traditionally, bone disease is diagnosed through the examination of X-ray pictures with the human eye of an experienced radiologist with limited success. Identifying indicators to detect hidden knowledge useful for solving a specified problem is generally difficult. However, a useful rule to reduce Nescience here is to increase accuracy of information by improving granularity in time and space. Image Biopsy Lab used this rule to develop a high-precision X-ray picture analysis of pixel granularity. This high-precision analysis reveals information hidden in the X-ray pictures of human bone. Each pixel shows a gray color level which is correlated to the mineral bone density. If the mineral bone density is lower than a necessary threshold value, the pixel signals a warning.

These case examples demonstrate how embracing Nescience can lead to groundbreaking discoveries and innovative solutions. By acknowledging what is unknown, these organizations applied a proactive and exploratory problem-solving approach, enabling them to overcome challenges, push the boundaries of knowledge, and revolutionize their respective fields.

6.8 Artificial Intelligence and Creative Problem-Solving

Artificial intelligence (AI) is the science and technology of using machines, such as computers and robots, to simulate human intelligence. Although still in its infancy, AI has recently advanced rapidly, especially in generative AI like ChatGPT and Google Bard for language and art simulation and in expert systems for large-scale analysis and human behavior predictions. We will explore the potential applications of AI to enhance creative problem-solving by first examine the special advantages and capabilities of AI and then identify potential application opportunities.

6.8.1 Special Advantages and Capabilities of AI

AI has the following special advantages:

- It is not susceptible to social control.
- It does not have psychological risk aversion.
- It does not have inherent cognitive rigidity.

Moreover, it has the following special capabilities:

It is highly trainable to learn new skills and adopt diverse perspectives and solution approaches.
It can have extensive knowledge bases.
It has super-fast computational speed.
It can work continuously without stop.
Artificial emotional intelligence for sensing, learning about, and interacting with human emotional life is under development, which can expand human cognition and awareness.

6.8.2 Applications of AI to Creative Problem-Solving

Based on these special advantages and capabilities, AI can be applied to creative problem-solving in many ways, including the following:

- *Provide diverse perspectives:* With its superb mimicking capability, AI can readily provide diverse perspectives based on different perceptions, expertise, and experiences.
- *Generate huge number of ideas:* With its super-fast computational speed, AI can be programmed to tirelessly generate huge number of ideas, creative or not.
- *Provide powerful search engine for past creative ideas from different fields:* Again, with its extensive knowledge base and super-fast computational speed, AI can easily develop a powerful search engine for finding and integrating past creative problem-solving ideas into a useful application knowledge base.

In sum, AI can not only enhance but will greatly expand and transform human capabilities for creative problem-solving.

6.9 Innovative Product Development and Technology Readiness Assessment

Innovative product development to creatively solve the problem arising from the needs and wants of the Final Adopter is a gradual and iterative process to reach Technology Readiness for the product to be presented to the market. This section presents the key concepts and processes for assessing Technology Readiness of the innovative product.

6.9.1 Key Elements of Technology Readiness

Technology Readiness consists of the following four key elements shown in Figure 6.4: Creative Problem-Solving Readiness, Intellectual Property Right Readiness, Integrability Readiness, and Production Readiness.

Figure 6.4 Key Elements of Technology Readiness

6.9.1.1 Creative Problem Solution Readiness

An innovative product development starts with a thorough understanding of the needs and wants of the Adopter that is market ready. It then forms an innovation team to generate and implement creative problem solutions demanded by the market need. Once a creative solution is tentatively selected, it goes through the stages of scientific, engineering, economic, ecologic, and equity feasibility tests, and then the proof of concept, prototype, minimal viable product, and scalability stages. Simultaneously, it will need to generate direct internal support and indirect customer acceptance. Only when the product has received final approval for test production, it reaches creative problem solution readiness.

6.9.1.2 Intellectual Property Rights Readiness

When an innovative product is under development, it needs intellectual property rights (IPR) protection. If an innovation team is convinced of the unique innovativeness of the problem solution, they can start patent application through the legal department of the organization.

In the United States, patentability is defined by 35 U.S.C. 101 Inventions Patentable: "Whoever invents or discovers any new and useful process, machine, manufacture, or composition of matter, or any new and useful improvement thereof, may obtain a patent therefore, subject to the conditions and requirements of this title."

Similar laws exist in other countries, but each must be individually applied through a professional IPR protection agency.

6.9.1.3 Integration Readiness

Next, the technology needs to be integrated as an innovative problem solution into the target customer system environment. Integration comprises the necessary and sufficient fulfillment of conditions that an additional element or subsystem is implanted in an existing operational target system. Hence the interfaces between the input and output of the added new system in communication with the corresponding input and output of the existing operational target system in terms of space, time, speed, content, and behavioral rules must follow the overall rules of system behavior.

6.9.1.4 Production Readiness

Finally, for an innovative product that is a physical good, it must be ready to be manufactured for mass production. The Production Readiness can be characterized by the configuration of production factors used in the production of the innovative product.

The basic categories of production factors are:

- *Materials*: Production material, auxiliary material (e.g., lubricants), operating materials (e.g., energy, vapor)
- *Tools*: Manual tools, machine tools, software tools
- *Labor*: Human labor, machine labor (semiautomatic or automatic), and robotic labor
- *Information*: List of parts, machine scheduling information, AI tools
- Type of production, including:
  - Fully manual production (e.g., manual wood carving)
  - Machine-supported manual production (e.g., sewing of shirts)
  - Concatenated computer numerically controlled (CNC) working machines production (e.g., multi-stage metal working with CNC machines)
  - Fully automated production (e.g., pharmaceutical products)
  - Robotic-supported small lot size production (welding of I-beams in oil rigs)

Moreover, due to the progressive development of digitization modularity of production systems as a design-driven strategy for being flexible and remaining effective when adapting to market-changing behavior has gained increasing importance. To cope with complexity, there are three application classes of modularity to be considered for Production Readiness [27].

- *Modularity in product design*: This modularity is mostly influenced by the design language and the creative ideas controlling the logical and physical design. Product design must cope with challenges of different design requirements. Functional material might be modified due to design requirements. For example, product design for use in harsh environments must balance the limits of functional materials with the extreme functional requirements (temperature, acid resistance, UV light resistance, etc.). Each coupling between modules can be considered as a risky exposure of functional intermodular communication.
- *Modularity in production*: This modularity considers the required degrees of freedom for production design optimization: degrees of automation,

customer-induced flexibility, tooling variety, and the resulting setup costs and times having a strong influence in small lot size production. The increasing customization and decreasing standardization have resulted in high aspiration levels for job scheduling, optimization of number of modules, and evolutionary design rules in manufacturing.

- *Modularity in organization*: This modularity is considered a key attribute supporting agility and flexibility in the intrapreneurial framework. Development of High-performance innovation team and the required communication and resource interfaces between the future innovation and existing production parts of the organization are critical challenges to the Intrapreneur Internal Supporter and the Final Adopter. Mutual respect and ethical communication are cultural requirements of effective and compliant intrapreneurship management.

On the other hand, for a product that is a service, a similar approach that substitutes scalability for production can be developed for Technology Readiness of production factors including tools (e.g., computer and communication equipment), information, labor, and standardization in place of modularity.

In sum, analysis of these key elements of Technology Readiness will result in a set of checklists and questionnaires to be input into the IRN © software presented in Chapter 8.

Summary

This chapter presents major barriers to creative problem-solving and proven approaches and tools for overcoming these barriers. It then presents the key elements of Technology Readiness for assessing the readiness of an innovative product to be introduced to the market for acceptance and adoption. Discussion of these key elements will be used as the basis to develop a set of checklists and questionnaires in Chapter 7 for the input of the Organization Readiness assessment as the starting part of the IRN © software. Again, it must be emphasized that the various creative problem-solving approaches and tools presented in this chapter are highly process-oriented and interactive, and their effective applications require expert-facilitated training and continuous practice on actual problems.

Glossary

Cognitive rigidity: The increasing rigidity in perception and thinking process due to indoctrination, reinforcement, and conservation of neurological energy as humans age, which is called psychological inertia by TRIZ.

Enneagram: A classification of nine personalities and steps for effective problem-solving.

Facilitated brainstorming: A popular approach to generating creative ideas through skilled facilitation that removes social control, eliminates risk aversion, and stimulates participation.

Nescience: The absence of knowledge which can be used for creative problem-solving by eliminating fixed mental models and perspectives, encouraging diverse thinking, and stimulating curiosity.

Risk aversion: The inherent asymmetry in human behavior to assign greater value to avoid the pain of failure rather than the pleasure of success.

Social control: The tendency for the authority in a group to impose conformity of thinking process for control and stability.

Traditional approach to creative problem-solving: The use of various traditional logical reasoning processes to develop creative problem solutions.

TRIZ: Russian acronym for "Theory of Inventive Problem-Solving" developed by Russian patent officer Genrich Altshuller and his followers.

Discussions

- Identify any additional barriers to creative problem-solving.
- Identify other approaches and tools for creative problem-solving.
- Critique the Enneagram approach for problem-solving by the Innovation Team.
- Critique TRIZ for creative problem-solving.
- Critique Nescience for creative problem-solving.
- Conduct a series of creative problem-solving exercises to experience the potential use of the various concepts and tools.
- Critique key elements of Technology Readiness.
- Develop an equivalent Manufacturability Readiness for a product that is a service.

References

1 Guilford, J. (1968). *Intelligence, Creativity, and Their Educational Implications*. Knapp.
2 Runco, M. (1990). Mindfulness and personal control [Review of Langer's mindfulness]. *Imagination, Cognition and Personality* 10: 107–114.
3 Szasz, T. (1988). *Pain and Pleasure, A Study of Bodily Feelings*. Syracuse University Press.
4 Shriver, A. (2014). The asymmetrical contributions of pleasure and pain to subjective well-being. *Review of Philosophy and Psychology* 5: 135–153.

5 Gopnik, A. and Schulz, L. (2007). *Causal Learning: Psychology, Philosophy, and Computation*. Oxford University Press.
6 Tenenbaum, J., Kemp, C., and Griffiths, T. (2006). Theory-based Bayesian models of inductive learning and reasoning. *Trends in Cognitive Sciences* 10 (7): 309–318.
7 Griffiths, T., Kemp, C., and Tenenbaum, J. (2008). Bayesian models of cognition. In: *The Cambridge Handbook of Computational Psychology*, 59–100. Cambridge University Press.
8 Sanborn, A. and Chater, N. (2016). Bayesian brains without probabilities. *Trends in Cognitive Sciences* 20 (12): 883–893.
9 Tobore, T. (2019). On energy efficiency and the brain's resistance to change: the neurological evolution of dogmatism and close-mindedness. *Psychological Reports* 122 (6): 2406–2416.
10 Gal, D. (2006). A psychological law of inertia and the illusion of loss aversion. *Judgment and Decision Making* 1 (1): 23–32.
11 Scholz, J., Klein, M., Behrens, T., and Johansen-Berg, H. (2009). Training induces changes in white-matter architecture. *Nature Neuroscience* 12 (11): 1370–1371.
12 Ericsson, K., Krampe, R., and Tesch-Römer, C. (1993). The role of deliberate practice in the acquisition of expert performance. *Psychological Review* 100 (3): 363–406.
13 Osborn, A. (1953). *Applied Imagination: Principles and Procedures of Creative Problem Solving*. Charles Scribner's Sons.
14 Tversky, A. and Kahneman, D. (1981). The framing of decisions and the psychology of choice. *Science, New Series* 211 (4481): 453–458.
15 Chevallier, A. (2016). *Strategic Thinking in Complex Problem Solving*. Oxford University Press.
16 Pham, C., Magistretti, S., and Dell'Era, C. (2023). How do you frame ill-defined problems? A study on creative logics in action. *Creativity, and Innovation Management* 32: 493–516.
17 Altschuller, G. (1996). *And Suddenly the Inventor Appeared: TRIZ, The Theory of Inventive Problem Solving*. Technical Innovation Center.
18 Sheu, D. (2019). *Mastering TRIZ Innovation Tools: Part I*, 5e. Agitek International Consulting, Inc. (In Chinese).
19 Velcro Corporate Website (2024). https://www.velcro.co.uk/original-thinking/our-story. 2024.
20 Fish, F., Howle, L., and Murray, M. (2008). Hydrodynamic flow control in marine mammals. *Integrative and Comparative Biology* 211: 1859–1867.
21 Schlegel, M. (2020). *Teamwork 9.0. Successful Workgroup Problem-Solving Using Enneagram*. Schlegel Publishing.
22 Sheu, D., Chiu, M., and Cayard, D. (2020). The 7 pillars of TRIZ philosophies. *Computers & Industrial Engineering* 146: 106572.

23 Sheu, D. (2019). *Systematic Patent Circumvention, Regeneration, and Enhancement*, 3e. Agitek International Consulting, Inc (In Chinese).

24 Sheu, D. (2019). *Mastering TRIZ Innovation Tools: Part II*, 5e. Agitek International Consulting, Inc. (In Chinese).

25 Weber, C., Hasenauer, R., and Mayande, N. (2017). Quantifying nescience: a decision aid for practicing managers. *Proceedings of PICMET Conference* (9–13 July 2017), Portland, OR, USA.

26 Chen, L. (2022). Cognitive flexibility and nescience: an empirical study. *Journal of Problem-Solving Research* 38 (4): 567–584.

27 Nilsen, J. (2003). Modularity and innovation. Working paper 03/2003, Institute pour Management dela Recherche et de l'Innovation. University of Paris Dauphine (June 2003).

7

Intrapreneurship Readiness Assessment: Methodology

Built on the concepts and discussions in Chapters 4–6, this chapter presents the methodology for assessing Intrapreneurship Readiness of an organization and an innovation project as the basis for the Intrapreneurship READINESS navigator (IRN©) software in Chapter 8.

7.1 General Methodological Approach for Readiness Assessment

The Intrapreneurship Readiness assessment process consists of two somewhat independent parts: Organization Readiness and Innovation Project Readiness through the iterative assessment of Market and Technology Readiness. An organization interested in Intrapreneurship can start Organization Readiness and continue with iterative Market and Technology assessment of an innovation project. Conversely, an aspiring intrapreneur can start with an iterative assessment of Market Readiness and Technology Readiness to guide an innovation project to success and leverage it to push for Organization Readiness through a grassroots effort. In the meantime, readiness assessment of an innovation project will depend on whether the project originates from market pull by a manifested demand for Adopter or a technology push by a creative breakthrough of an Innovation Team. Each assessment will use three major ingredients: values, information, and decision rules, to develop a set of readiness questionnaires/checklists. The degree of readiness is based on the extent of values achieved, the sufficiency of information acquired, and the validity of the decision rules.

The intrapreneurship readiness assessment is complex because, as discussed in Chapter 2, the innovation process is interactive, dynamic, and evolving.

Intrapreneurship Management: Concepts, Methods, and Software for Managing Technological Innovation in Organizations, First Edition. Rainer Hasenauer and Oliver Yu.

Published 2024 by John Wiley & Sons, Inc.

Moreover, there are many different kinds of innovative products including both goods and services, as well as management methods. Thus, it is impossible to have a finite set of assessment processes covering all possible market and technology interactions and evolutions. Moreover, the assessment is a learning process for the Assessment Team who are either Internal Supporters or Innovation Team. As the team iterates between Market and Technology Readiness of an innovation project, they learn how to absorb new knowledge, experiences, and insights to improve the assessment process. As a result, we will outline a general approach with sample checklists that have been proven useful in past assessments by the coauthors for assessing overall Organization Readiness as well as the iterative Market and Technology Readiness of over 100 innovation projects. The general approach and sample checklists can then be expanded and customized for specific applications by the Assessment Team through the interactive and adaptive IRN© software presented in Chapter 8. Moreover, we will also provide additional examples of the specific assessment processes used in the case studies in Appendix A for Organization Readiness and Appendix B for Market and Technology Readiness.

It is important to point out that, as discussed extensively in Chapter 3, achieving values for all innovation participants: Innovation Team, Internal Supporter, and Adopter, is the ultimate goal for Intrapreneurship Management. As a result, all readiness checklists must include meeting the integrated Economic–Ecologic–Equity (EEE) values as the fundamental requirement for both an organization and a project. This requirement is especially important and relevant with the increasing public awareness of the global circular economy [1] and the growing demand for corporate compliance with the UN Sustainable Development Goals.

In implementing the general approach, the Organization Readiness assessment is comparatively more straightforward and algorithmic. The Assessment Team starts with a set of initial sample checklists on the readiness of an organization in terms of Innovative Culture, Internal Support, and Innovation Team. These initial checklists can be expanded based on discussions and interactions among Assessment Team members. The levels of overall Organization Readiness will be assigned based on a simple algorithm that can also be modified and customized by the team as it gains experience and insights.

Implementing the approach for the iterative Market and Technology Readiness assessment of an innovation project is more complex because of the interactions between Market and Technology as well as the complexity of the actual adoption of the product by the Adopter. Specifically, even if the project has reached both Market and Technology Readiness, the resulting product may still not be adopted because it has not met the marketability criteria as discussed in Chapter 5. As a result, the assessment starts with respective sample master checklists for Market and

Technology Readiness, supported by their respective set of sub-checklists for the readiness of the key sub-elements. All these checklists can be modified, expanded, and customized as the Assessment Team gains experience and insights into the process. Moreover, the team needs to determine at the beginning whether the innovation project is driven mainly by market pull or technology push. If it is driven by market pull, where the Market Readiness is generally high, the technology development needs clear specifications to meet the market needs. If the project is driven by a technology push, then it starts with no clearly understood market needs. In both cases, it is important to use the Market Test Bed (MTB), discussed in Chapter 5, to apply marketability criteria to establish requirements for market entry based on the adaptability of the product.

There are two other important considerations for the readiness assessment process. First, the assessment is a repeating process. Thus, it will be carried out based on "milestones" reached, which can be a point in time or the occurrence of a significant event in organization evolution or project development. Second, although the sample checklists presented in this chapter for Market and Technology Readiness assessments appear largely for hardware innovations, the general principles are applicable to software, service, and management method innovations.

The general relationships among the key elements of Intrapreneurship Readiness assessment are outlined in Figure 7.1.

In the following, we will discuss more details of the three readiness assessments.

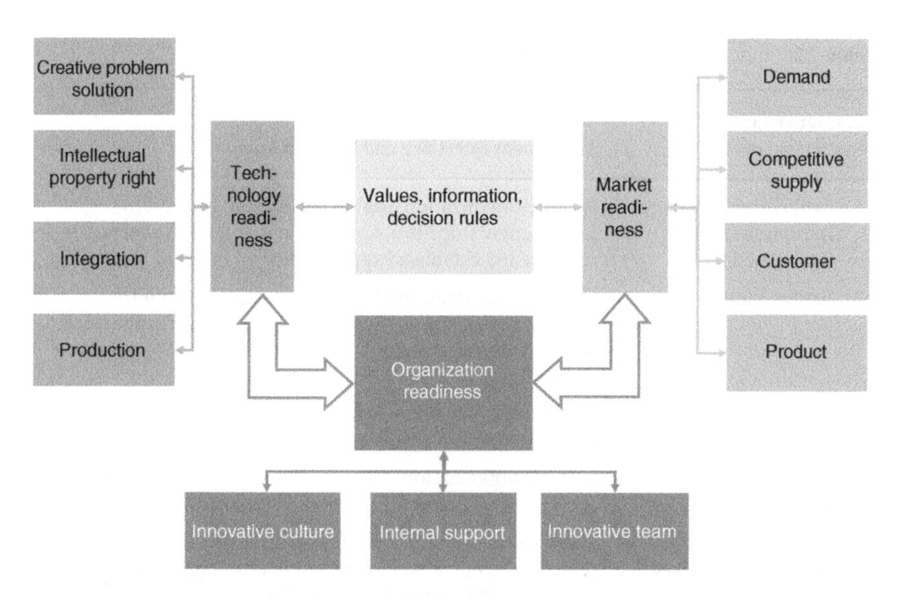

Figure 7.1 Relationship Among Key Elements of Intrapreneurship Readiness Assessment

7.2 Organization Readiness

As discussed in Chapter 4, Organization Readiness consists of three key elements: Innovative Culture Readiness, Internal Support Readiness, and Innovation Team Readiness. These three types of readiness are sequentially related. Specifically, if the assessment is conducted by an organization interested in initiating Intrapreneurship from the top, it would start with Innovative Culture Readiness, then Internal Support Readiness, and finally Innovation Team Readiness. On the other hand, if the assessment is initiated by an aspiring intrapreneur wanting to build Intrapreneurship through a grassroots effort, it would start with Innovation Team Readiness to ensure a strong team, then Internal Support Readiness to attract Internal Support, and finally use Innovative Culture assessment to promote an Innovative Culture for the organization.

7.2.1 Innovative Culture Readiness

The following is a sample checklist for this key factor of Organization Readiness to assess the readiness for the development and promotion of an Innovative Culture in an organization (Table 7.1).

For this checklist, Levels 2–7 can be iterative and interactive. Generally, an organization is ready to start innovation projects after reaching Level 7 or 70% of the questionnaire and will attract internal innovators when it reaches Level 10.

Table 7.1 Innovative Culture Readiness Checklist

Innovative Culture Readiness Level	Assessment Questions and Decision Rules
1) Explicit interest in Intrapreneurship established	Has the organization established explicit interest in Intrapreneurship? If not, review and amend the corporate mission and establish explicit policy on internal innovation.
2) Values aligned	Has the organization aligned its values through the integrated Economic–Ecologic–Equity value framework for both the organization and the workers? If not, align the values.
3) Financial resources allocated	Has the organization allocated financial resources for Intrapreneurship? If not, allocate.
4) Incentive programs developed	Has the organization developed incentive programs for Intrapreneurship at all employee levels? If not, develop one.
5) General support processes developed	Has the organization developed support processes, such as project development and innovation training programs, to support Intrapreneurship? If not, develop one.

Table 7.1 (Continued)

Innovative Culture Readiness Level	Assessment Questions and Decision Rules
6) Attractive work environment developed	Has the organization developed attractive physical and enriching social environment to promote Intrapreneurship? If not, develop one.
7) Project management process established	Has the organization established effective project management processes and training programs? If not, establish the process.
8) Internal innovation project history	Has the organization had a history of innovation projects? If yes, assess the results; if no, start a project.
9) Innovation image promoted	Has the organization promoted its image of being innovative? If not, start promotion.
10) Reputation established	Has the organization established a reputation for being innovative? If not, establish a reputation.

7.2.2 Internal Support Readiness

The following is a sample checklist for this key factor of Organization Readiness to assess the readiness of the organization to provide detailed specific support to internal innovation (Table 7.2).

With the Internal Supporter being the interface between the organization and Innovation Team, Internal Support Readiness strongly affects Innovative Culture Readiness. Most of the levels are interactive and can be concurrent. Specifically, Levels 1, 5, and 8 are general organizational support and the other levels are individual support by Internal Supporter. These levels can all be achieved simultaneously.

It cannot be overemphasized that to reach high levels of Organization Readiness, many organizational development tools like Agile, Scrum, and Enneagram discussed in Chapter 4 must be professionally applied and practiced with expert-facilitated training in an organization.

7.2.3 Innovation Team Readiness

The following is a sample checklist for this key factor of Organization Readiness to assess the readiness of Innovation Team for internal innovation (Table 7.3).

The first five levels can be concurrent. However, Level 6 on Team Champion is crucial to the readiness of this key factor of Organization Readiness, as "No Champion, No Project" has been the proven rule for ensuring innovation project

Table 7.2 Internal Support Readiness Checklist

Internal Support Readiness Level	Assessment Questions and Decision Rules
1) Internal Supporter identified and provided	Does the organization have an established process to identify and provide an Internal Supporter as a mentor and interface with corporate management? If not, establish the process.
2) Values aligned and a common understanding of Internal Innovation reached	Are Internal Supporter, corporate management, and Innovation Team aligned in the integrated EEE value framework, and have they reached a common understanding of the goal and process of internal innovation? If not, align values and develop common understanding.
3) Competence and mutual respect possessed	Does Internal Supporter possess sufficient competence in communication and management to achieve mutual respect with the Innovation Team to support internal innovation? If not, consider a more competent and respected Internal Supporter.
4) Internal connections developed	Is Internal Supporter well connected with, and can draw resources from other parts of the organization? If not, improve connections.
5) Project assessment and decision process established	Is there a rigorous process for assessing a new idea and its implementation plan and the subsequent decision-making that has a track record of success? If not, establish one.
6) Corporate communications developed	Has Internal Supporter developed a collaborative interface with corporate management and Innovation Team for effective communication and reporting of agreed key performance indicators? If not, establish the interface.
7) Internal Supporter evolutionary capability developed	Has the Internal Supporter developed the ability to evolve effectively with changing goals and environment? If not, learn to develop the capability. If not, develop the capability.
8) Celebration of learning established	Does the organization have a process to celebrate the learning from any failure of an innovation project? If not, develop one to provide positive support and encouragement to internal innovation. If not, establish the process.

success. Furthermore, an effective Team Champion can generally ensure the team will reach Levels 7–9 successfully.

Finally, past experiences by the authors have shown that an organization is ready for internal innovation if each of the three key elements reaches at least 70% of its readiness level or questionnaire with positive answers.

Table 7.3 Innovation Team Readiness Checklist

Innovation Team Readiness Level	Assessment Questions and Decision Rules
1) Common understanding achieved	Does the team have a common understanding of the goal and process of Intrapreneurship? If not, achieve common understanding.
2) Values aligned	Are the values of the team members aligned with the integrated EEE value framework? If not, align values.
3) Roles defined	Are the roles of the team members well-defined and complementary? If not, define roles.
4) Team diversity acquired	Does the team have sufficient diversity in including at least a technologist and a marketer? If not, acquire or develop diversity in team member perspectives and competence.
5) Mutual understanding and respect established	Do the team members understand and respect the strengths, weaknesses, major roles, and personality styles of each other? If not, take training to establish empathy and mutual respect.
6) Team champion acquired	Does the team have a champion who is committed to the success of the innovation project? If not, acquire a dedicated champion who is critical to the project's success.
7) Supporter interaction developed	Has constant productive interaction been developed with Internal Supporter? If not, develop close interactions.
8) Adopter collaboration established	Has constant collaboration with the Final Adopter been developed? If not, establish collaboration.
9) evolutionary capability	Has the team developed the capability to evolve effectively with changing goals and environments? If not, develop capability.

7.3 Market Readiness

Market Readiness is a key element of Innovation Project Readiness. It is the starting point of the iterative assessment with Technology Readiness if the project is initiated through market pull of existing demand of Adopter. On the other hand, if the project is initiated by a creative breakthrough of the Innovation Team, then it generally has a low Market Readiness at the beginning and requires additional research and resources to reach a sufficient readiness level. A sample checklist for overall Market Readiness of an innovation project that has been successfully applied in the past is shown below with detailed descriptions (Table 7.4).

This master checklist can be supplemented, enhanced, modified, and expanded through the insights from the following sample readiness checklists of the key elements or factors of Market Readiness discussed in Chapter 5.

Table 7.4 Market Readiness Master Checklist

Market Readiness Level	Detailed Description and Decision Rules
1) Unsatisfied needs have been identified	A fuzzy feeling of latent needs to solve a perceived problem and identification of subtle, weak signals. Decision: Verify feeling. If valid, identify potential early Adopter and intensify market research; otherwise, keep observing and undertake no action to avoid "White Elephant."
2) Identification of specific needs	Deeper understanding of latent/emerging needs achieved and specific needs identified. Decision: Discuss benefit expectations and constraints for the need with early Adopter. If not clear, intensify market research and improve communication.
3) Identification of the functionalities for new products (goods or services)	Describe functional content to cover the need for new products or services and identify specific expectations and constraints for the expressed needs of identified early Adopter. If not clear, intensify understanding of expected user benefits and user preferences.
4) Quantification of expected functionalities	Develop a model of expected functionalities and effective Adopter's constraints and expected Adopter's behavioral features as a data source for future production, marketing, and sales strategy.
5) Identification of system capabilities	Set up a definition of system capabilities to deliver valid responses to identified needs covering product/service design, identify effective suppliers for purchasing tangible and nontangible components; set up quality design and production design with required capacity for the planning horizon.
6) Translation of the expected functionalities into needed capabilities to build the response	Based on the given system capabilities of previous MR level, the expected product/service functionality items are translated into the required quantitative and qualitative product units/service units and in compliance with the expected product/service design composed together to the functional prototype being response ready for early Adopter evaluation.
7) Definition of the necessary and sufficient competencies and resources	Taking early Adopter's feedback into account, the design of productive system machinery, list of human workshops, quality control operations as well as packaging, delivery, and invoicing processes are planned and ready to be offered as a response to the market.
8) Identification of the experts possessing the competencies	For a sustainable and profitable business, availability of production and logistics as well as administrative support with desired and economically feasible list of machines, workers, and staff must be financed.
9) Response to the expressed need in the market	Finally, the management board will decide whether to respond to the need in market by offering an adequate set of assorted product lines according to the market segmentation plan in compliance with different customer segments.

7.3.1 Demand Readiness

Demand Readiness assesses the state of demand for the innovative product and the shift in willingness to use and pay along the path to market entry. The work of Florin Paun [2] and one of the authors of this book [3] has been used to develop a set of indicators for progressive levels of readiness along the path to market entry based on the innovation marketing strategy. These indicators along the path have been developed through market observation and direct contact with early prospects to provide validity of Demand Readiness in terms of willingness to use and to pay (Table 7.5).

Table 7.5 Demand Readiness Checklist

Demand Readiness Level	Detailed Description and Decision Rules
1) Occurrence of feeling "something is missing"	Feeling "something is missing" indicates a latent demand, which is the start of a market. If it occurs, describe the need based on the feeling.
2) Identification of specific need	All attributes of the specific need are well-defined and are operationally documented. If not, special market studies and time/cost budget analysis are required.
3) Identification of expected attributes for new product	Expected functionalities to meet the latent/emergent need are identified through backward engineering process with special emphasis on EEE values. If not, complete the identification process.
4) Quantification of expected functionality	The identified functionalities and their required attributes are expressed in quantified and operational scale units. If not, Innovation Team needs to conduct market research and prototype tests with support of internal/external experts for the quantification.
5) Identification of system capabilities	Required system capabilities for system environment, input, output, states, architecture (degree of modularity, hierarchy), process architecture (multi/concurrent tasking), behavior (deterministic, stochastic, time/space dependency), and integrability (interfacing and coupling) and future-proof design are identified. If not, identify.
6) Translation of expected functionality into needed abilities to respond to market	A backward engineering process translates the expected functionalities into required resources and capabilities to build the problem-solving response with supply chain risk analysis and EEE value requirements. If not, complete functionality report with risk estimation.
7) Identification of necessary and sufficient competencies and required resources	Necessary and sufficient knowledge, skills, and license (insourcing of patent rights, specification of future EEE value requirements) for market and business development as well as marketing measures are dedicated resources and are identified and documented. If not, do it.

(Continued)

Table 7.5 (Continued)

Demand Readiness Level	Detailed Description and Decision Rules
8) Identification of experts possessing the competencies	Knowing the resource requirements enables focused search and acquisition of expert manpower with a selective and qualified human resource acquisition process. Market research, market communication, market segmentation, price models, financial terms, supply chain design, and customer relationship management are the main chapters of functional responsibility of the hired experts.
9) Building initial answer to the expressed need in the market	Use resources, together with Product Readiness, develop initial answers to the well-defined need, and communicate it to the target market segment. If not, develop market communication strategy in compliance with corporation's communication strategy.

7.3.2 Competitive Supply Readiness

Competitive Supply Readiness assesses, positions, and tracks the innovative product competitiveness against the strongest competitor known to the Innovation Team and determines the threshold to estimate Innovation Half-Life. The following sample checklist can be used by the Innovation Team to position its own competitive strength and track the path to market entry (Table 7.6).

Past experiences have shown that a Competitive Supply Readiness Level below 6 may result in the postponement of the development of the competitive plan

Table 7.6 Competitive Supply Readiness Checklist

Competitive Supply Readiness Level	Detailed Description and Decision Rules
1) Information sources for target market research identified	Research and information collection on how and source of "something is missing" is completed, and other complementary and substitutional offers in the market are checked. If not, do it.
2) Data sources for competitor observation identified	Data sources for players/stakeholders in the market value chain and their interaction (aims, positioning, roles, cooperation, etc.) and potential synergies and own market opportunities are identified. If not, collect and document data sources.
3) First ranking of important direct competitors done	Important direct competitors are ranked with their own differentiation and market value chain interactions. If not, develop unique selling propositions (USPs) and conduct competitive validation study.

Table 7.6 (Continued)

Competitive Supply Readiness Level	Detailed Description and Decision Rules
4) Intellectual property rights (IPR) comparison ranking done	Own differentiation and market value chain interaction and strategy based on IPR and EEE values are done. If not, conduct IPR comparison ranking.
5) Technological and functional comparison ranking done	First market feedback with players/stakeholders in the market value chain is established. If not, urge the market for feedback to motivate the project team.
6) Competitive target product physically evaluated	Relations with players/stakeholders in the market value chain are established and the first estimate of Innovation Half-Life of the strongest competitor's product is done with validation tests. If not, do them.
7) Competitive product-market fit evaluated	First tests are done and potential benefits for interaction are confirmed. If not, be careful about breakeven time to market.
8) Innovation Half-Life evaluated	Values to and from players/stakeholders in market value chain are evaluated and marketing strategy against competitor is ready to avoid risk of losing time and acceptance in the market.
9) Plan for proactive countermeasures elaborated	Strategic coopetition options are formulated, and strategy is developed with a business model for collateral effects. If not, set up strategy team to develop plan.
10) Overall competitive scenario available	Competitive strategy for selected scenarios with Innovation Team its Championis done. If not, risk losing the innovation game.

until the urgent need to study the collateral effects on Demand Readiness and Customer Readiness has been fulfilled.

7.3.3 Customer Readiness

Customer Readiness assesses the preparedness and willingness of the customer to proactively show interest and be ready to listen and invest time and intellectual resources to become familiar with the announced innovation. Deep and valid understanding is the precondition for understanding and trusting the innovative solution and the intrapreneur as supplier.

In business-to-business (B2B) marketing, customer organization is frequently represented by *a buying decision group* consisting of people with different roles and expertise, including purchasing manager (price and paying conditions), quality assurance/control specialists quality key performance indicators (KPIs) as well as MTBF (Mean Time Between Failure) and MTTR (Mean Time To Repair), production manager (integrability, setup time, setup costs), maintenance manager (MTBF, spare part availability, early warning signals, expected downtime and restart frequency, tied up capital with special tooling), and legal professionals (Table 7.7).

Table 7.7 Customer Readiness Checklist

Customer Readiness Level	Detailed Description and Decision Rules
1) Hypothesis on possible needs in market developed	Latent/emergent needs with weak signals are identified by Innovation Team through creative hypothesis making. If not, use observation and creativity to generate hypotheses.
2) Identified specific needs in market	Specific needs by exploratory interviews and focus groups are identified with description of needs by the attributes of missing functionality, typical consequences of malfunctioning, direct and collateral damages, failure propagation effects, etc. If not, do them.
3) First market feedback established	Based on qualitative research with problem-centered interviews, focus groups, and explorative studies, a first study of market feedback is established with quantitative indicators of market potential.
4) Confirmed needs from several customers and/or users	Needs from potential customers/users of competitive products are confirmed with documentation of the operational features of the malfunctioning or missing solution, and an operational description of the expected innovative solution and the benefit components. Helpful tools include functional prototypes like augmented reality and simulation models to convey the proposed solution to early prospects. If not, set up project experts to expand knowledge.
5) Established relations with target customers	Relations with early prospects willing to cooperate on the development of the prototype or to be a member of a community-based innovation (CBI) are established. If not, find the subset of prospects and establish relations.
6) Benefits of the product confirmed by partnerships or first customer testing	Benefits of development and test partnership from CBI, and acceptance test of business model are confirmed for testing willingness to use and to pay. If not, set up an experienced team for testing perceived willingness to use and to pay, in compliance with corporation goals.
7) Customers in extended product testing and/or first test sales	First test sales along the first supply chain and first customer satisfaction are documented with reports of first failure management, workaround practices, spare part delivery time, and setup of MTBF and MTTR reports. If not, do it to develop trust and gain experience with customer demand.

Table 7.7 (Continued)

Customer Readiness Level	Detailed Description and Decision Rules
8) Information collection system on first products sold set up	Sales information systems for all stages from acquisition to product installation, training, maintenance, spare part operations, supply chain satisfaction, and end customer satisfaction reports are set up. If not, customer relations will suffer serious quality losses in competence and trust. Work out countermeasures.
9) Widespread product sales	Sales system is working, and monthly sales system report is established for profitability, growth, market share, and EEE values.
10) Strategic value and ranking established	After sustained market entry, strategic value and ranking of the innovative product in fulfilling the corporation goals through the EEE value evaluation is established.

7.3.4 Product Readiness

Product Readiness levels list serves to finalize product innovation marketing toward addressed market segments and required product certification standards in compliance with corporation communication standards. Is the product ready for widespread use in compliance with the regulatory requirements of the respective market authorities and their related co-authorities? This is an important question for the Innovation Team. The following is a sample checklist for Product Readiness as a way to answer this question.

Product Readiness in case of widespread use is strongly influenced by regional and geopolitical standardization law and by international trade usage and course of dealing. At this stage of preparing market entry for widespread use, Innovation Teams' experience will be enriched by additional mixed teams with corporation experience and external trade usage experience. The challenge for champion's role is to balance all these constraints and opportunities without a minimum risk of missing the window of opportunity for market entry and smart control of market entry budget (Table 7.8).

7.4 Technology Readiness

Technology Readiness master checklist used here is based on a set of widely applied concepts for invention and innovation tracking [5, 6]. Although Technology Readiness has a long history of being analyzed by itself, in actual applications, due to economic considerations and intrapreneurial challenges for an innovative

Table 7.8 Product Readiness Checklist

Product Readiness Level	Detailed Description and Decision Rules
1) Target market identified	Based on market research, first target market is identified with competitive scenery, main prospects customers, market regulations, and prevailing business models. If not, identify.
2) Markets segments and lead users' needs defined, and competing products analyzed	Valid segmentation criteria selected, segmentation executed, lead users identified, competitors and their relevant products identified, market potential per segment evaluated, and results checked. If not, do all these.
3) Plan for product options and extended product family formulated	Based on segmentation results and required differentiation of benefit per segment, the development of product family or product line strategy and its related production and marketing parameters for segment-specific market communication is ready. If not, do it.
4) Marketing plan developed	Business model developed with marketing plan including distribution strategy (agents, supply chain fees, distribution contracts, value-added policy), communication strategy, price and payment policy, customer relationship management, customer satisfaction reporting, etc., and compatible with EEE values. If not, do it according to Joyce, et al [4].
5) Promotion and launch materials developed	Design, testing, and production of promotion tools, printed materials, social media network policy, blogging, content management, exhibition and trade fairs plan, and reporting of marketing key performance indicator as basis for marketing controlling, developed with selective selling strategy focusing segments which are most promising to achieve EEE goals.
6) Field testing facilitated	Field tests, mystery shopping plans, documentation of quality feedback, and customer/prospects feedback have been set up. Perceived quality, ease of use, and usefulness of product documented. Price test and willingness to pay and to use done. Augmented reality is used for experimental testing.
7) Regulatory approval/ certification obtained	All documents of approval for the target market and future target markets are ready, and all certifications obtained.
8) Early production ramp-up products placed with preferred customers. Active service and support secured	Start production and delivery routines with early customers, develop proactive service, support, and training/spare part and maintenance service provision. If not, set up a priority plan by reducing complexity and focusing on the most promising customer segment considering EEE goals.
9) Product promotion and market entry ready	Start promotion activities, market entry plans, and control.
10) Quality assurance of comprehensive Product Readiness	Set up EEE-compatible product documentation.

solution to gain market entry dictate that Technology Readiness assessment must constantly iterate with Market Readiness assessment to account for the interplay between supply and demand [7] (Table 7.9).

7.4.1 Creative Problem Solution Readiness

Technology Readiness starts with the understanding of the needs and wants of Adopter, and then the application of the creative solution techniques to identify creative problem solutions. It is to be followed by a series of feasibility studies and various test product developments of these potential solutions to generate Internal Support and external acceptance before the potential solutions receive approval for test productions (Table 7.10).

7.4.2 Intellectual Property Rights (IPR) Readiness

The USP of an innovative product often requires the protection of IPR. Aiming at a strong market entry, the Innovation Team needs to analyze how IPR Readiness can be achieved, which results in a set of criteria for IPR Readiness [8] (Table 7.11).

Table 7.9 Technology Readiness Master Checklist

Technology Readiness Level	Detailed Description and Decision Rules
1) Fundamental research	Conduct required fundamental research to validate technology feasibility.
2) Applied research	Establish features of necessary and sufficient conditions for applied research for the technology.
3) Research to prove feasibility	Set up, execute, and report feasibility studies.
4) Laboratory demonstration	Set up, execute, and report laboratory demonstrations.
5) Technology development	Decide, design, apply, and control efficiency parameters.
6) Whole system field demonstration	Determine technology efficiency through whole system field demonstration.
7) Industrial prototype	Develop industrial prototype for envisaged market.
8) Product industrialization	Set up industrialization design and execution.
9) Market/sales certification	Apply for all necessary market/sales certifications.

Table 7.10 Creative Problem Solution Readiness Checklist

Creative Problem Solution (CPS) Readiness Level	Description
1) Adopter needs and wants are understood	Intrapreneurs must first understand the needs and wants of Adopter through observation and empathy.
2) Creative problem-solving techniques applied	Innovation Team must apply various creative problem-solving techniques to identify potential solutions to the problem of satisfying the needs and wants of Adopter.
3) Scientific and engineering feasibility studies conducted	For potential solutions, studies will be conducted to test scientific and engineering feasibility.
4) Economic–Ecologic–Equity (EEE) values assessed	For creative solutions with scientific and engineering feasibility, the integrated EEE values will be assessed.
5) Proof of concept developed	For creative solutions with high EEE values, proof of concept will be developed.
6) Prototype developed	For solution with promising proof of concept, prototype will be developed.
7) Minimal viable product (MVP) developed	For the most promising solutions, MVP will be developed.
8) Understanding of Adopter needs and EEE value reassessed	For creative solutions with MVPs, understanding of the Adopter needs and wants and EEE values will be reassessed to ensure Adopter acceptance and Internal Support.
9) Test production approved	The creative solutions receive organizational approval for test production.

Table 7.11 Intellectual Property Rights Readiness Checklist

IPR Readiness Level	Detailed Descriptions and Decision Rules
1) Hypothesizing on possible IPR (patentable inventions)	Hypothesizing possible patents or some other form of IPR in results or ideas. A vague description of the IPR and what is unique. Some ideas for patenting, etc. may exist but are speculative. Limited or nonexistent knowledge of the technical field, state of art, publications, etc.
2) Identified specific patentable inventions or other IPR	Familiarity with the technical field, state of art, and publications within the field. Specific patent ideas exist but are unvalidated and not necessarily derived by commercial considerations. Agreements related to IPR are identified and ownership is clarified and verified for IPR control.

Table 7.11 (Continued)

IPR Readiness Level	Detailed Descriptions and Decision Rules
3) Detailed description of possible patentable inventions. Initial search of the technical field and prior art	Sufficiently detailed description of possible inventions (according to template). Some description of other forms of possible IPR. Discussion/analysis is made by professionals of what is patentable and what other forms of IPR exist in the project. Made own searches/analysis of prior art in the field? Possible initial searches by professionals to find prior art done?
4) Confirmed novelty and patentability; Decided on alternative IP protection if not patenting	Confirmed novelty through searches by professionals. Confirmed patentability by analysis by a professional. Possibly filed one or several provisional applications i.e., not professionally drafted and not complex with claims, etc. If patents are not considered suitable (after professional analysis) decide on possible alternative forms of IP protection.
5) First complete patent application filed, draft of IPR strategy done	First complete patent application is filed in cooperation with a professional (selected for the field/business). Patent strategy: Professional analysis on what/how to patent and how to improve/build the value of patent application (supporting data, new/additional details to be filed, etc.). Draft IPR strategy: First analysis, preferably supported by professional, e.g., on how different IPR can be used. Basic agreements are put into place to ascertain control of IPR (e.g., assignments, ownership of copyright, etc.)
6) Positive response on patent application; Initial assessment of freedom to operate (FTO), patent strategy supporting business	Positive response on patent application received from authorities (national and international) and analysis of response performed. If not, conduct professional analysis with strong arguments and strategy for patent prosecution. Develop professionally validated IPR strategy in support of business strategy and identify additional patents, country strategy, possible claim changes, etc. Initial assessment of FTO, e.g., competitor-based, narrowed product scope, etc.
7) Patent entry into national phase; Other formal IPR registered	Entry into national phase (US, EU, Japan, etc.). Complementary additional new patents might be filed. Other forms of IPR might be registered such as FTO, trademarks, designs.
8) First patent granted, IPR strategy fully implemented, more complete assessment of FTO	First patent is granted with relevant scope for business. No opposition was encountered for patent grant. IPR strategy is fully implemented. IPR is proactively used to support business, e.g., all IPR-related arguments are professionally managed. More complete assessment of FTO.
9) Patent granted in relevant countries, strong IPR support for business	Patent granted in several countries relevant for business. Patent is maintained in force. Patent is evaluated to provide business value. Patent is in force/valid with no invalidation procedures. Since IPR supports and protects business, for example using various other forms or registered IPR (trademarks, designs, etc.) or for example using agreements, trade secrets, etc.

7.4.3 Integration Readiness

Integration addresses the necessary and sufficient conditions for an innovative product to be integrated into an existing operational target system. Hence the interfaces between the input and output of the additional new system communicating with the corresponding output and input of the existing operational target system in terms of space, time, speed, content, and behavioral rules must be in compliance with the overall rules of target system behavior (Table 7.12).

Integration Readiness is key for successful implementation of the innovative product in the target customer environment. Poor integrability should be the basis

Table 7.12 Integration Readiness Checklist

Integration Readiness Level	Detailed Descriptions and Decision Rules
1) Interface between technologies identified with sufficient detail to allow characterization of the relationship	This is the lowest level of integration readiness and describes the selection of a medium for integration.
2) Some level of specificity to characterize the interaction between technologies through their respective interface	Once a medium has been defined, a "signaling" method must be selected such that two integrating technologies are able to influence each other over that medium. As a result, this level represents the proof of concept for integration.
3) There is compatibility (i.e., common language) between technologies to orderly and efficiently integrate and interact	This is the minimum required level for successful integration. At this level, two technologies can not only influence each other but also communicate interpretable data. Many technology integrations have failed this level due to the incorrect assumption that if two technologies can exchange information successfully, then they are fully integrated.
4) Sufficient detail in the quality and assurance of the integration between technologies	This level goes beyond simple data exchange and requires that data sent are the data received with verification. If not, Innovation Team must redesign their data concept to fulfill this requirement.
5) Sufficient control between technologies necessary to establish, manage, and terminate the integration	This level means that two or more of the integrating technologies can control the integration itself through establishing, maintaining, and terminating the integration.

Table 7.12 (Continued)

Integration Readiness Level	Detailed Descriptions and Decision Rules
6) The integrating technologies can accept, translate, and structure information for its intended application	The highest technical level to be achieved, which includes the ability to control integration and what information to exchange, unit labels to specify what the information is, and the ability to translate from a foreign data structure to a local one.
7) The integration of technologies has been verified and validated with sufficient detail to be actionable	This represents a significant step beyond the previous level in that in addition to a technical perspective, the integration also works from a requirements perspective such as performance, throughput, reliability, and availability.
8) Actual integration completed and mission qualified through test and demonstration, in the system environment	Integration meets requirements, with a system level demonstration in the relevant environment, revealing defects not discovered until the interaction of the integrating technologies is observed in the system environment.
9) Integration is mission proven through successful mission operations	Integrated technologies are used in the system environment successfully. For a technology to move to final Technology Readiness, it must first be integrated into the system, and then proven in the relevant environment. Thus, achieving this level implies that the component technology has matured to final Technology Readiness.

for stopping product development and redesigning integrability. Communicative integration means that communication protocol not only confirms sending and receiving signals on lowest Open System Interconnection (OSI) layer but also sending/receiving messages with intended content without loss of meaning and intention by coding and decoding.

7.4.4 Production Readiness

Producing an innovative product is a key activity, risky but decisive to gain competitive advantage in terms of superior functionality, material cost, labor advantages, and a profitable and compliant business model for growing and convincing the envisaged market [9, 10]. The following is the checklist of Production Readiness for a product that is good. For a product that is a service, a similar checklist can be constructed (Table 7.13).

Table 7.13 Production Readiness Checklist

Production Readiness Level	Detailed Descriptions and Decision Rules
1) Basic production implications identified	The lowest level of Production Readiness, focusing on production bottlenecks and weak links for understanding how to avoid unexpected production breakdown. Consider redundancy in energy sources, uninterrupted power supply (UPS), and critical resources in staff, machines, and IT.
2) Production concepts identified	Level 2 is characterized by describing the application of new production concepts. Applied research translates basic research into solutions for broadly defined industrial needs. Typically, this level includes identification, paper studies, and analysis of material and process approaches. An understanding of production feasibility and risk is emerging. These studies consider also the EEE values and goals.
3) Production proof of concept developed	Level 3 begins the validation of the production concepts through analytical or laboratory experiments. This level is typical of technologies in the funding of applied R&D. Materials and/or processes have been characterized for producibility and availability, but further evaluation and demonstration are required. Experimental hardware models have been developed in a laboratory environment that may possess limited functionality. Production design should be in compliance with EEE values.
4) Capability to produce the technology in a laboratory environment	Level 4 is typical for R&D programs and their respective budgets and acts as an exit criterion for the material solution analysis phase decision milestone. At this level, required investments for production technology development have been identified. Processes to ensure manufacturability, producibility, and quality are in place and sufficient to produce technology demonstrators. Production risks have been identified for building prototypes and mitigation plans are in place. Producibility assessments of design concepts have been completed. Key design performance parameters have been identified as well as any special tooling, facilities, and material handling skills required. EEE values help reduce material costs and achieve environmental goals.
5) Capability to produce prototype components in a production-relevant environment	Level 5 is typical of the mid-point in the technology development phase of acquisition, or in the case of key technologies, near the mid-point of an advanced technology demonstration project. The industrial base has been assessed to identify potential production sources. A production strategy has been refined and integrated with the risk management plan. Identification of enabling/critical technologies and components is complete. Prototype materials, tooling and test equipment, and personnel skills have been demonstrated on components in a production-relevant environment, but many production processes and procedures are still in development. Production technology development efforts have been initiated or are ongoing. Producibility assessments of key technologies and components are ongoing. A cost model has been constructed to assess projected production costs. EEE values are integral to production design.

Table 7.13 (Continued)

Production Readiness Level	Detailed Descriptions and Decision Rules
6) Capability to produce a prototype system or subsystem in a production-relevant environment	Level 6 is normally at R&D completion and acceptance into a preliminary system design. Initial production approach has been developed. Most production processes have been defined and characterized, but there are still significant engineering and/or design changes in the system itself. However, preliminary design of critical components has been completed and producibility assessments of key technologies are complete. Prototype materials, tooling, and test equipment, as well as personnel skills, have been demonstrated on systems and/or subsystems in a production-relevant environment. Cost analysis has been performed to assess projected cost versus target cost objectives and the program has in place appropriate risk reduction to achieve cost requirements or establish a new baseline. Long-lead and key supply chain elements were identified and EEE values were achieved.
7) Capability to produce systems, subsystems, or components in a production-representative environment	Level 7 is typical for system detailed design activity that is underway. Material specifications have been approved and materials are available to meet the planned pilot line build schedule. Production processes and procedures have been demonstrated in a production-representative environment. Detailed producibility studies and risk assessments are underway. The cost model has been updated with detailed designs, rolled up to system level, and tracked against allocated EEE goals. Unit cost reduction efforts have been prioritized and are underway. The supply chain and supplier quality assurance have been assessed and long-lead procurement plans are in place. Production tooling and test equipment design and development have been initiated.
8) Pilot line capability demonstrated; Ready to begin low-rate initial production	Level 8 is associated with readiness to enter into low-rate initial production. Technologies should have matured to at least TRL 7. Detailed system design is essentially complete and sufficiently stable to enter low-rate production. All materials are available to meet the planned low-rate production schedule. Production and quality processes and procedures have been proven in a pilot line environment and are under control and ready for low-rate production. Known producibility risk poses no significant challenges for low-rate production. The engineering cost model is driven by detailed design and has been validated with actual data. The industrial capabilities assessment has been completed and shows that the supply chain is established and stable.

(*Continued*)

Table 7.13 (Continued)

Production Readiness Level	Detailed Descriptions and Decision Rules
9) Low-rate production demonstrated; Capability in place to begin full-rate production	Level 9 is when the system, component, or item has been previously produced, is in production, or has successfully achieved low-rate initial production. This level of readiness is normally associated with readiness for entry into full-rate production. All systems engineering/design requirements should have been met such that there are minimal system changes. Major system design features are stable and have been proven in tests. Materials are available to meet planned rate production schedules. Manufacturing process capability in a low-rate production environment is at an appropriate quality level to meet design key characteristic tolerances. Production risk monitoring is ongoing. Cost targets have been met, learning curves have been analyzed with actual data. The cost model has been developed for full-rate production environment and reflects the impact of continuous improvement.
10) Full-rate production demonstrated and lean production practices in place	Level 10 is when production is normally associated with the production or sustainment phases of the acquisition life cycle. Engineering/design changes are few and generally limited to quality and cost improvements. System, components, or items are in full-rate production and meet all engineering, performance, quality, and reliability requirements. Production process capability is at the appropriate quality level. All materials, tooling, inspection, test equipment, facilities, and manpower are in place and have met full-rate production requirements. Rate production unit costs meet goals, and funding is sufficient for production at required rates. Lean practices and circular processes are well-established and continuous process improvements are ongoing.

7.5 Special Application of the Readiness Assessment Process

As presented in Appendix A, Dr. Curtis Carlson, the former president and CEO of SRI International, has developed and instituted a highly successful innovation project development process through the use of value creation forum, where an Innovation Team presents to a group of intelligent and experienced technical and business managers and staff members, every two weeks, a 5- to 10-minute presentation on the specific market needs to be met by the project, the creative solution approach used to meet the need, the expected benefit to cost of the prospective product, and the results of the competitive analysis (NABC) and receive constructive criticisms from the audience. The Market and Technology Readiness assessments can be effectively applied to improve the contents and readiness of the innovation project during the two weeks between successive presentations.

Summary

This chapter outlines a general approach and presents a collection of sample checklists for assessing the Organization Readiness of an organization for Intrapreneurship, and the Market and Technology Readiness of an internal innovation product development and implementation in an organization. These sample checklists can be expanded, modified, and customized as the Assessment Team gain experience and insights from the assessment. The degree of readiness is dependent on the extent of values achieved, the sufficiency of information acquired, and the validity of the decision rules. There is a particular strong emphasis on the integrated EEE values for both organizations and projects alike, especially for a sustainable circular global economy.

Glossary

Milestone: Either a preset point in time or the time of a major achievement in organization evolution or project development that a new readiness assessment is needed.

Readiness checklist: A set of criteria or questionnaires for checking the progressive levels of readiness of an organization or object to reach a specified goal; measurement of readiness is done by checking if the criteria used to describe the level are fulfilled.

Readiness assessment process: A process for the members of an Assessment Team to assess the level of readiness of an organization or object to reach a specified goal.

Discussions

- Discuss and potentially improve the general approach for Organization, Market, and Technology Readiness assessments.
- Discuss and potentially improve the set of sample checklists for the assessments.

References

1 (2019). EU project. www.circularstart.eu. Cooperation for innovation and the exchange of good practice in sustainability and circularity training of startups.

2 Paun, F. (2011). Demand Readiness Level (DRL), a new tool to hybridize market pull and technology push approaches. *Proceedings of A NR-ERANET Workshop* (February 2011), Paris, France.

3 Hasenauer, R., Gschöpf, A., and Weber, C. (2016). Technology readiness, market readiness and the triple bottom line: an empirical analysis of innovating startups in an incubator. *Proceedings of PICMET Conference*, September 4–8, 2016, Honolulu, Hawaii, USA, pp. 1387–1428.

4 Joyce, A., Paquin, R., and Pigneur, Y. (2015). The triple layered business model canvas: a tool to design more sustainable business models, *Proceedings of ARTEM Organizational Creativity International Conference,* May 2015, Nancy, France.

5 Hasenauer, R., Weber, C., Filo, P., and Orgonáš, J. (2015). Managing technology push through marketing testbeds: the case of the Hi-Tech Center in Vienna, Austria. *Proceedings of PICMET Conference* (August 2–6, 2015), Portland, Oregon, USA, pp. 99–126.

6 U.S. Department of Energy (2011). Technology readiness assessment guide DOE G 413.3-4A 9-15-2011. www.directives.doe.gov.

7 Mankins, J. (1995). *Technology Readiness Levels, (White Paper).* Advanced Concepts Office, Office of Space Access and Technology, NASA (6 April 1995).

8 Isoz, A., Coordinator (2014). KTH innovation: innovation and collaboration: self-evaluation report on innovation support process. *Proceedings of AAE* (2 March 2014).

9 OSD Manufacturing Readiness Level (MRL) (2011). Deskbook version 2.0, prepared by the OSD manufacturing technology program in collaboration with The Joint Service/Industry MRL Working Group and updated version (May 2011).

10 OSD Manufacturing Technology Program (2020). Manufacturing readiness level (MRL) deskbook version. In collaboration with the Joint Service/Industry MRL Working Group.

8

Intrapreneurship Readiness Assessment: Software Outline

This chapter provides an outline of the general structures, contents, and applications of Intrapreneurship READINESS navigator (IRN©) software, which is a member of the READINESS navigator (RN©) software developed by ONTEC AG, Vienna, Austria. The innovation project readiness assessment part of Intrapreneurship Management for iterative Market Readiness and Technology Readiness assessments has been developed and has operated as a special application of the RN© software for over 15 years. The new IRN© software will now add the Organization Readiness module. For the completeness of the book, we present here the outline of a provisional version of the IRN© software that is already functional through the existing RN© software. A final version of the IRN© software with additional customizations will be available with the publication of the book. The following description of the provisional version of the IRN© software is provided by Christian Rathgeber, the chief architect of the RN© software.

8.1 Overall Structure and Key Functions

The IRN© software provides a web-based platform that works like a very detailed, intelligent assessment sheet that can be individually adapted for different user and target groups. Additionally, creative ideas, innovations, products, or even organizations can be compared with industry indices. This enables early detection of areas for innovation improvement for an organization based on its Organization Readiness Level (ORL) for Intrapreneurship Management as well as critical deviations from the desired Technology Readiness Level (TRL) and Market Readiness Level (MRL) of an innovation project. In addition, much more than a mere

Intrapreneurship Management: Concepts, Methods, and Software for Managing Technological Innovation in Organizations, First Edition. Rainer Hasenauer and Oliver Yu.

Published 2024 by John Wiley & Sons, Inc.

inventory, the software suggests possible solutions and next steps depending on the level of detail of the input. For example, if a company wants to start an Intrapreneurship program, the IRN© software will ask whether a corporate mission statement has been developed and suggest sample recommendations. Similarly, if an innovation team wants to focus its efforts on the manufacturing process as part of the development of a new product, the IRN© software will ask, among other things, whether a patent has already been registered and suggest prioritizing this step.

Depending on the degree of maturity of the company, the IRN© software will provide appropriate indicators relevant to the impact, so that the corporate policies and programs can be modified and enhanced while the company continues its normal operations, thereby significantly increasing its efficiency. Similarly, with IRN© software being applied to an innovation project, the project can be controlled, counteracted, or corrected while it is still running, thereby greatly increasing its probability of success. The solution thus supports proactive decisions and measures through the early visualization of risk positions in corporate or project performance, based on known industry benchmarks. It is important to note that the authors have deliberately made the structure of the software adaptive and easily modifiable, so that it is not only capable of learning and using what it has learned from the user to evaluate and improve other corporate programs and different innovation projects but also can incorporate future advances in Intrapreneurship Management concepts and tools.

For the provisional version of IRN© software, we will use the existing nomenclature and structure of the current RN© software, in that an assessment is called a "project" and the registered user is called the "project owner."

The following are the key functions of the IRN© software.

8.1.1 Input Master and Other Readiness Checklists

The Assessment Team Leader is the registered user or project owner in the provisional version. Other Assessment Team members need to be invited by the team leader to be project owners.

The registered user or project owner can enter a set of master checklists for Organization Readiness, and Market and Technology Readiness with respectively associated sub-checklists into the system or change or supplement existing checklists.

8.1.2 Continue Master Readiness Checklists

The master checklists continue until changed by the Administrator or Assessment Team, i.e., the project owner in the current provisional version of IRN© software.

8.1.3 Select Module

The IRN© software contains two modules: Intrapreneurship Management-Organization Readiness and Intrapreneurship Management-Market and Technology Readiness.

In the current provisional version of IRN© software imbibed in the RN© software, the selection of the module will be in the form of a "project." The Assessment Team can select the module and its associated checklists by entering the project as either "Intrapreneurship Readiness Navigator/ORL" or "Intrapreneurship Readiness Navigator/MRL & TRL." The Assessment Team can use, expand, or modify the checklists in each module.

In the course of the use of a module, three default milestones "project start," "first milestone," and "project end" are created to market the timeline for the application of the software and its iterations.

After completion of assessment with one of the modules, the user is led directly to the maintenance of the milestones.

8.1.4 Maintain Assessment Master Data

The Assessment Team can change all assessment master data at any time. However, existing rating data for all checklists continue to be maintained for historical records and possible future use.

8.1.5 Send Invitation

The team leader of the Assessment Team can invite other people to collaborate/coevaluate the project or organization. These people do not have to be known in the system, in which case the email address of the invitee must be entered. The invitee will receive an email. If the addressee is not yet known in the system, he/she will receive a link to register this email.

8.1.6 Accept Invitation

The person invited by the Assessment Team can decline or accept the invitation. If the invitation is accepted, the module will appear on the module page. The invitee can see all data in the module but cannot change it; likewise, the milestones cannot be changed by the invitee.

8.1.7 Maintain Milestones

The assessment is made for each active milestone, which can be a periodic time, significant progress, or the achievement of a level or readiness. At the beginning of the assessment, the Assessment Team can define the time or the development cycle and identify the individual development phases or stages with milestones. The lead time in weeks to reach the milestone is specified for each milestone. As long as a milestone has not yet been accepted, it can be moved/reordered in the order of the not yet accepted milestones. In order to record a review, there must be at least one accepted milestone that has not yet been reached. Marking a milestone as "achieved" makes the next accepted milestone the active one.

8.1.8 Personal Rating

Members of the Assessment Team and each of the people invited can give personal assessments at an active milestone.

8.1.9 Official Rating

Only the Assessment Team can submit an official readiness assessment at an active milestone.

8.1.10 Calculate Level

Generally, the lowest level achieved in a sub-checklist acts as the readiness level for the master checklist. Other algorithms, as well as group discussions, may be used as the basis for calculating the readiness for the master checklist as well as recommendations for action to improve the readiness level in the master checklist.

8.2 User Interface

By navigating to the IRN's URL, a person gets to a welcome screen, where he/she can register as a user in the IRN ("create account") or in the header of which already registered users can log in to the IRN.

8.2.1 Welcome Page/Login

The user will first turn on the Welcome Page and be asked to enter a valid combination of "username" and "password."

8.2.2 New Account

If the user does not have an account, he/she will go to "new account" to register for a new account.

8.2.3 Forgot Password

If the user forgets the password, he/she will be redirected to the "reset password" page. By clicking the "reset password" button, a mail containing a link to the password change can be requested.

8.2.4 The IRN Software Module

Once logged in, the user will be directed to a screen showing two choices: "Intrapreneurship/ORL" for Organization Readiness assessment, or "Intrapreneurship Management/MRL & TRL" for Market and Technology Readiness assessment of an innovation project.

The following are input screens for ORL and MRL & TRL, respectively (Figure 8.1).

Assessment Name

This field holds the assessment name.

Assessment Status

This field has the initial value "active."

Possible values are:

- *Active*: Active assessment in an organization or project development cycle.
- *Inactive*: Assessment in which no measurable further development has taken place over a long period of time.
- *Completed*: Assessment has successfully completed the organization or project development cycle.
- *On Hold*: Assessment has been halted for the time being.
- *Dead*: Assessment was canceled or discontinued.

Origin Industry

This field contains the code of the industry segment in which the organization or project is located.

Target Industry

This field contains the code of the industry segment for which the organization or project is being developed.

Objective

Short definition of the organization or project goal.

Description

More details on the organization characteristics or project content, the organization or project environment, etc.

Checklists Included

A set of checklists has been included. The Assessment Team can switch off individual questions, groups of questions, sub-checklists, or even the entire master checklist and thus mark them as not applicable to the organization or project. However, at least one assessment criterion must remain active.

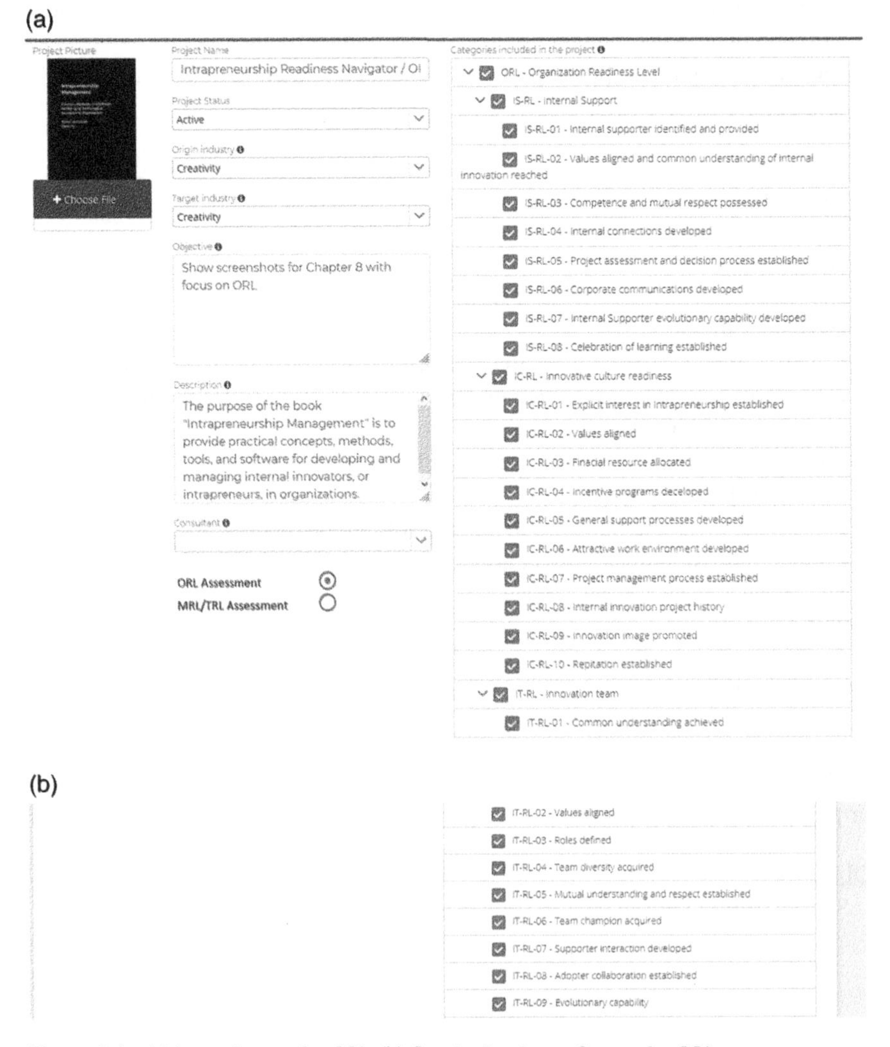

Figure 8.1 (a) Input Screen for ORL. (b) Continuing Input Screen for ORL

(c)

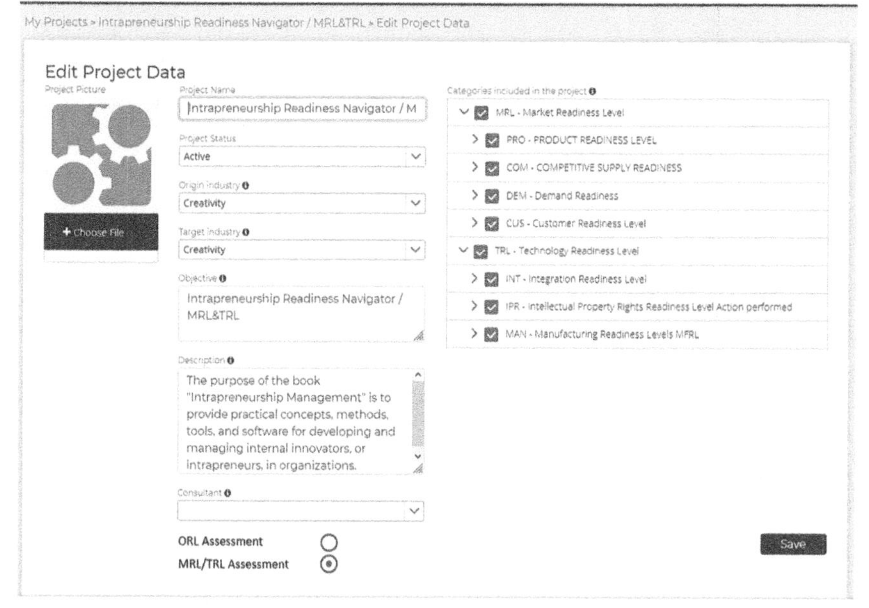

Figure 8.1 (c) Input Screen for MRL & TRL

8.2.5 Assessment Page

The assessment page is the central element of the user interface for the provisional IRN© software. From here, milestones can be edited, ratings submitted, organization or project master data maintained, users invited, required skills/resources specified, and comments entered.

Only those buttons are displayed for which the user is authorized on the one hand and which are permitted in the current status on the other.

Edit Milestones Schedule

By using milestones, the iterations of an assessment can be tracked on a timeline (Figure 8.2).

Milestones are those points in time in the project at which certain intermediate goals should or must be achieved. At these points in time, the assessment for the individual readiness levels is also carried out. Milestones that have already been completed cannot be changed afterward.

In the course of creating a new assessment, the system creates three initial milestones as defaults and proposes them to the user for further processing. However, if required, the user can define his own milestones using the "add new milestone" button. Milestones can be changed using the "edit milestone" button as long as they are "gray," i.e., not yet accepted. A milestone element is activated by clicking on the milestone and moving to the right in the graphic representation.

(a)

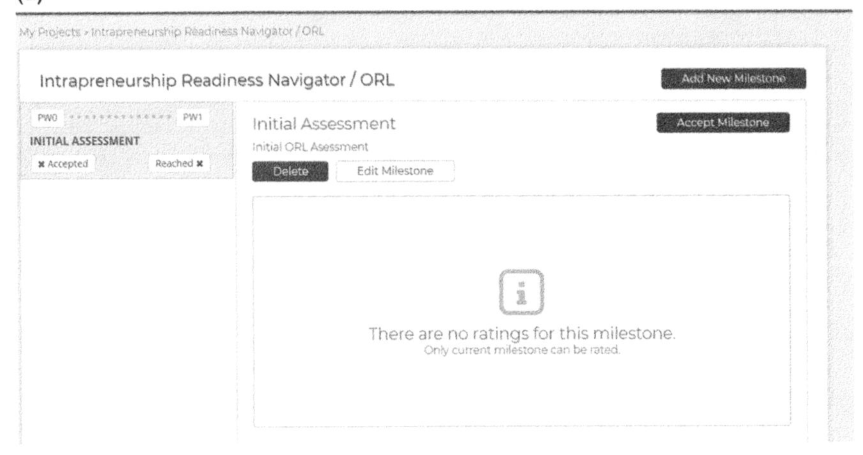

(b)

Figure 8.2 (a) Edit Milestone Schedule for ORL. (b) Edit Milestone Schedule for MRL & TRL

8.2.6 Rating at a Milestone

As soon as at least one milestone has been accepted, an assessment is possible. As long as this milestone is not marked as reached, an individual (My Rating) of all participants as well as an official rating by the Assessment Team is possible.

Recommended procedure for the assessment: In a so-called "planning poker" [1], each team member, i.e., each person invited to the assessment, submits his/her personal rating. The individual assessments are discussed in a joint meeting and, at the end, agreement is reached on the official assessment.

When the milestone is reached, the current assessments are recorded and cannot be changed. The current milestone is shown with a yellow/orange background, and the name of the milestone and its description are displayed in the assessment area.

Only the Assessment Team can see the "official rating" and "my rating" tabs; invited participants can only see the "my ratings" tab. The tab with a white background is the currently active one, and the inactive one has a gray background.

On the right edge of the browser window, there are buttons for possible additional actions on the project (Figure 8.3).

Show All Ratings

After clicking on this button, all assessments available for the milestone are displayed in a table. A click on the arrow next to the individual rating entry opens this rating tree, up to sub-checklist level at most.

Button "Pie Chart"

After clicking on this button, the display of the evaluation diagram changes from bar format to circle format. The levels of the sub-checklists are displayed in circle format.

Rate "xxx"

By clicking on the rating button you go to the rating mask for ORL, MRL or TRL.

Note: Text About the Rating Level

By clicking on the assessment bar of a checklist, this area is activated, and an explanatory description of the level achieved together with the determined potential for improvement of the overall assessment of the subject of this checklist.

If a link between two checklists is defined in the master checklist, a button for displaying these two checklists in a two-dimensional diagram, including the development of the assessment from the previous milestones, is displayed below the information text. This diagram is displayed below the hint text.

Rate

By clicking on the "rate" button of the respective main category, you branch into the assessment scheme of this checklist.

The following are rate screens for ORL and MRL & TRL (Figure 8.4).

The respective sub-readiness is selected by clicking on the designation field, and the background color of the designation field changes from blue to white. The total number of assessment levels of this subcategory is displayed in the designation field, preceded by the currently recorded level.

As an example, in the subcategory "Manufacturing Readiness" of the main category "Technology Readiness," the current rating is 3 out of a total of 10 rating levels in this subcategory.

The individual stages of the assessment level must be activated or deactivated sequentially by clicking, it is not possible to skip an evaluation level (in both directions).

(a)

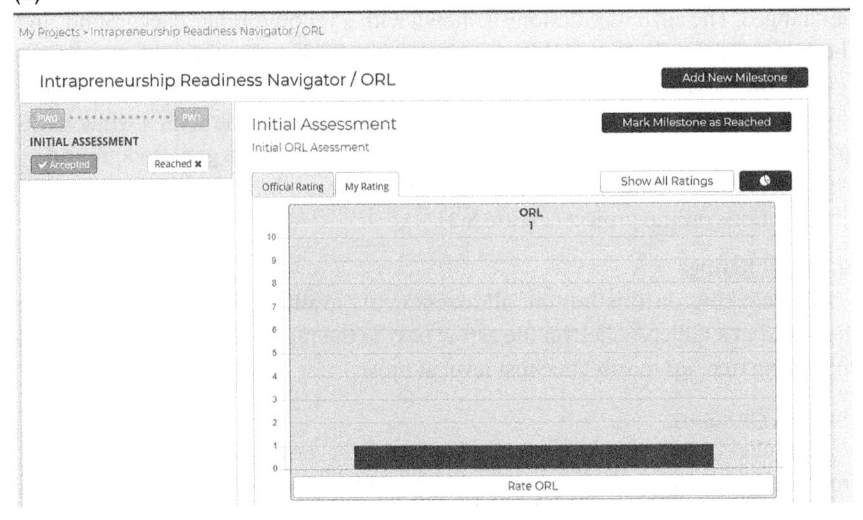

(b)

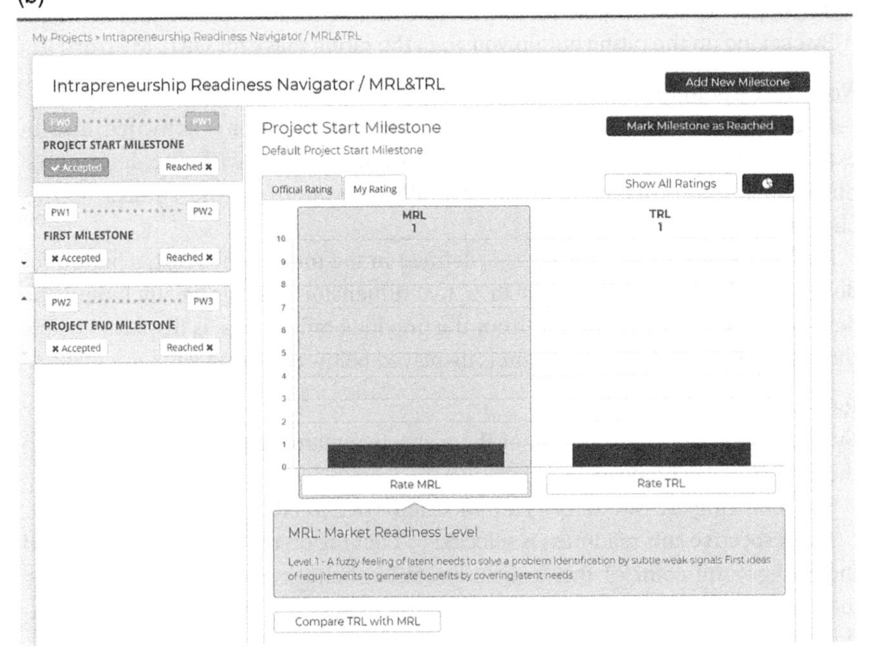

Figure 8.3 (a) Rating at a Milestone for ORL. (b) Rating at a Milestone for MRL & TRL

(a)

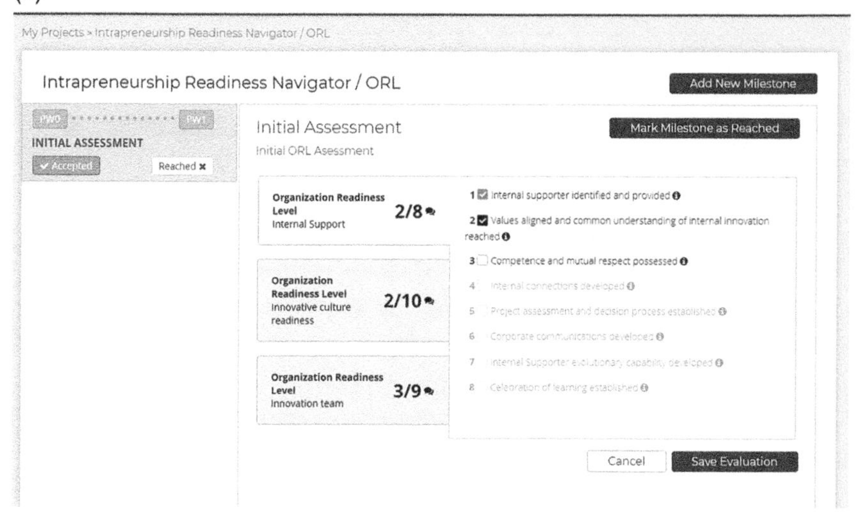

(b)

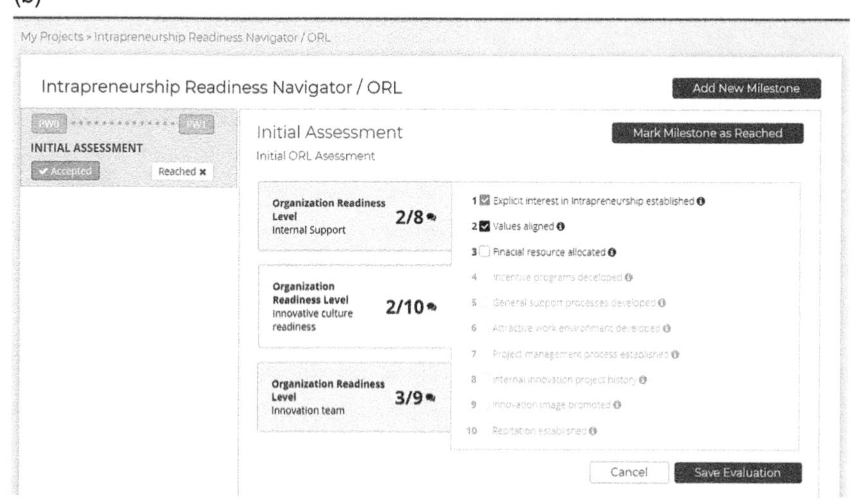

Figure 8.4 (a) Rate Screen for ORL (I). (b) Rate Screen for ORL (II)

(c)

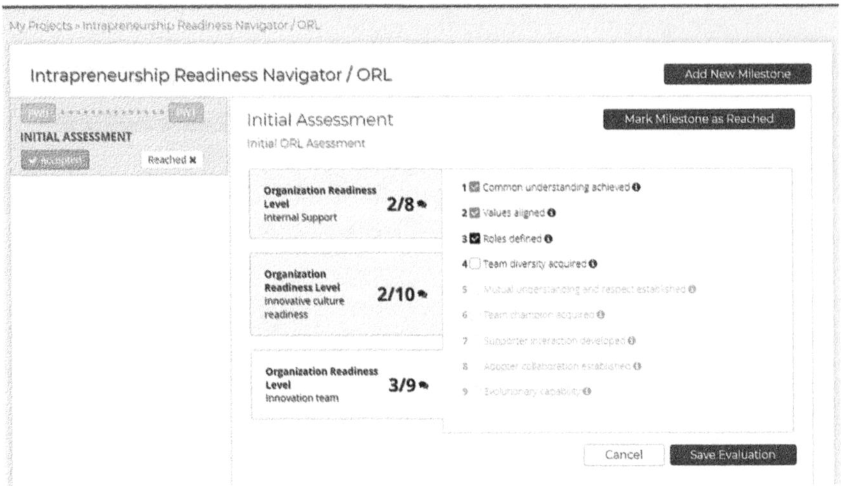

My Projects › Intrapreneurship Readiness Navigator / ORL

Intrapreneurship Readiness Navigator / ORL

Add New Milestone

INITIAL ASSESSMENT

Accepted

Reached ✖

Initial Assessment

Initial ORL Assessment

Mark Milestone as Reached

Organization Readiness Level
Internal Support
2/8

Organization Readiness Level
Innovative culture readiness
2/10

Organization Readiness Level
Innovation team
3/9

1 Common understanding achieved
2 Values aligned
3 Roles defined
4 Team diversity acquired
5 Mutual understanding and respect established
6 Team champion acquired
7 Supporter interaction developed
8 Adopter collaboration established
9 Evolutionary capability

Cancel

Save Evaluation

(d)

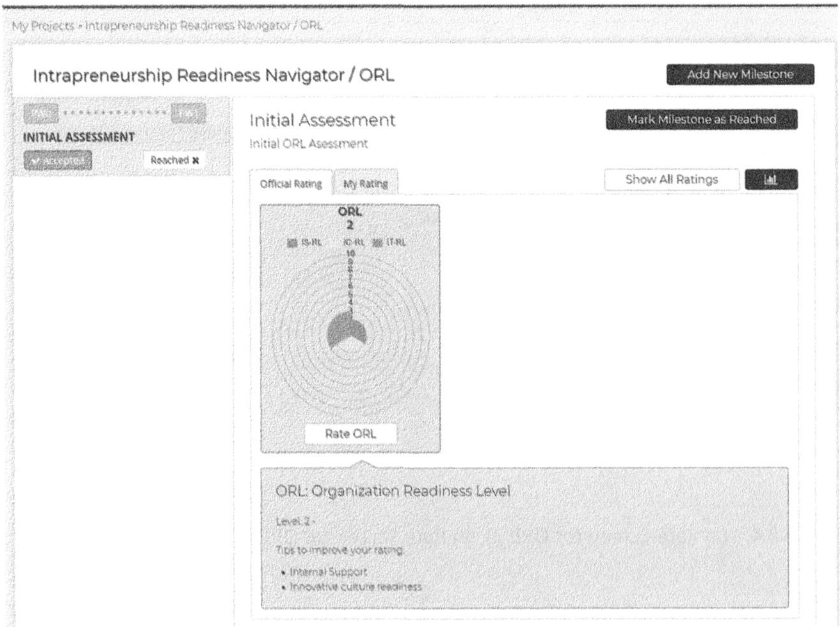

My Projects › Intrapreneurship Readiness Navigator / ORL

Intrapreneurship Readiness Navigator / ORL

Add New Milestone

INITIAL ASSESSMENT

Accepted

Reached ✖

Initial Assessment

Initial ORL Assessment

Mark Milestone as Reached

Official Rating

My Rating

Show All Ratings

ORL
2

IS-RL
IC-RL
IT-RL

Rate ORL

ORL: Organization Readiness Level

Level 2 -

Tips to improve your rating:

- Internal Support
- Innovative culture readiness

Figure 8.4 (c) Rate Screen for ORL (III). (d) Rate ORL Results

Figure 8.4 (e) All Ratings for ORL. (f) Rate Screen for MRL

(g)

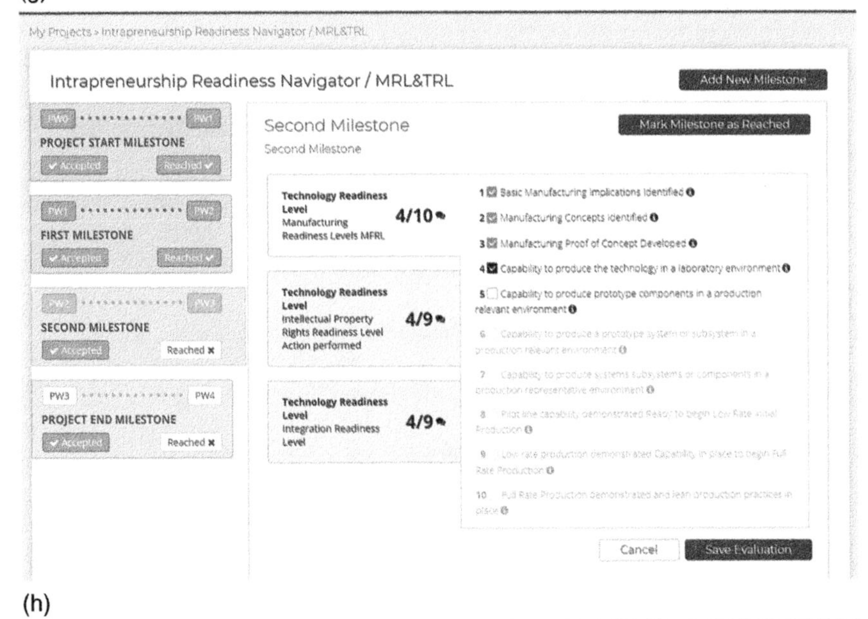

(h)

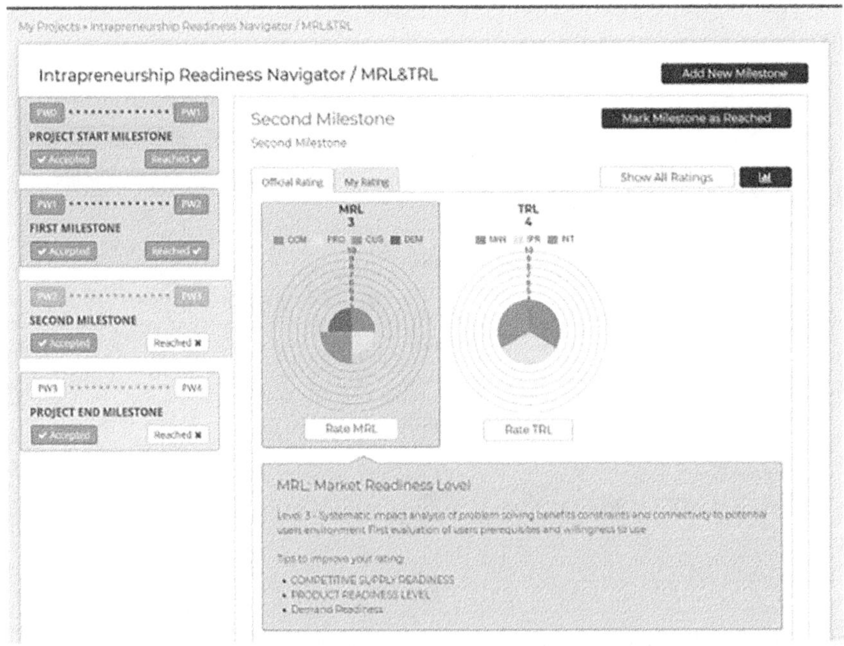

Figure 8.4 (g) Rate Screen for TRL. (h) Rate Screen for MRL&TRL Results

(i)

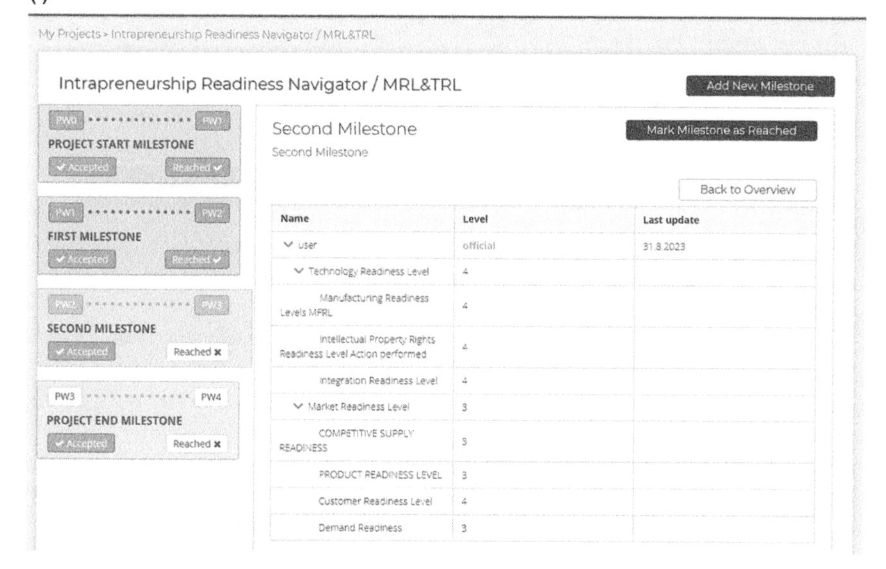

Figure 8.4 (i) All Ratings for MRL & TRL

In addition to the individual rating levels, you can position the mouse pointer on the information symbol ⓘ and a short explanatory text about the level will be displayed.

By clicking on the comment symbol, a comment on the rating of the subcategory can be entered in the designation field, and existing comments are displayed (Figure 8.5).

8.2.7 Assessment Menu

By clicking on the "menu" button, an overview of the assessment data such as Assessment Team members, comments on the assessment, and registered skills are displayed. Buttons for capturing invitations and skills are only visible to the Assessment Team, while comments can be captured by all project participants. The menu slides in from the right with a semitransparent background (Figure 8.6).

Invite/Invite User

The Assessment Team can invite others to participate in the assessment. This is done either by nominating an existing user with valid ID, or by providing the invitee's email address. In the latter case, the invited person receives an email invitation to register.

After clicking on the "invite" or "invite user" button, a window for entering the user to be invited is displayed. Existing users can be invited by selecting/entering their user Id, but for people who are not yet known in the system, their email

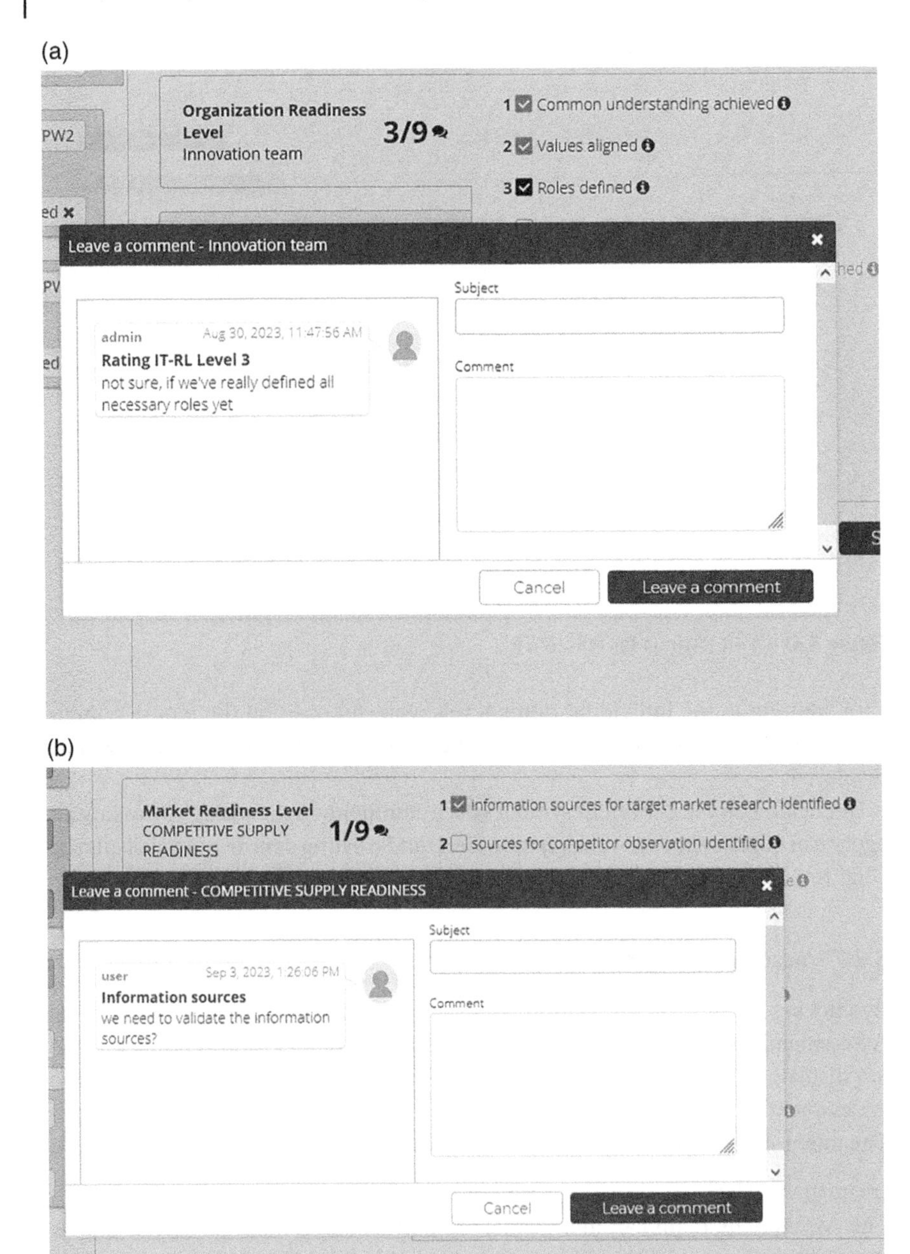

Figure 8.5 (a) Rating Comments for ORL. (b) Rating Comments MRL

(c)

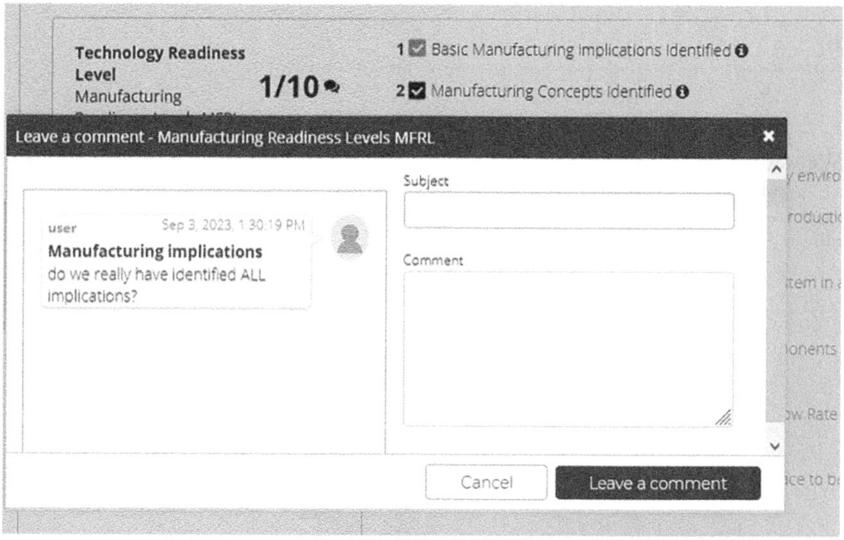

Technology Readiness Level
Manufacturing
1/10
1 Basic Manufacturing implications Identified
2 Manufacturing Concepts Identified
Leave a comment - Manufacturing Readiness Levels MFRL
user
Sep 3, 2023, 1:30:19 PM
Manufacturing implications
do we really have identified ALL implications?
Subject
Comment
Cancel
Leave a comment

Figure 8.5 (c) Rating Comments TRL

(a)

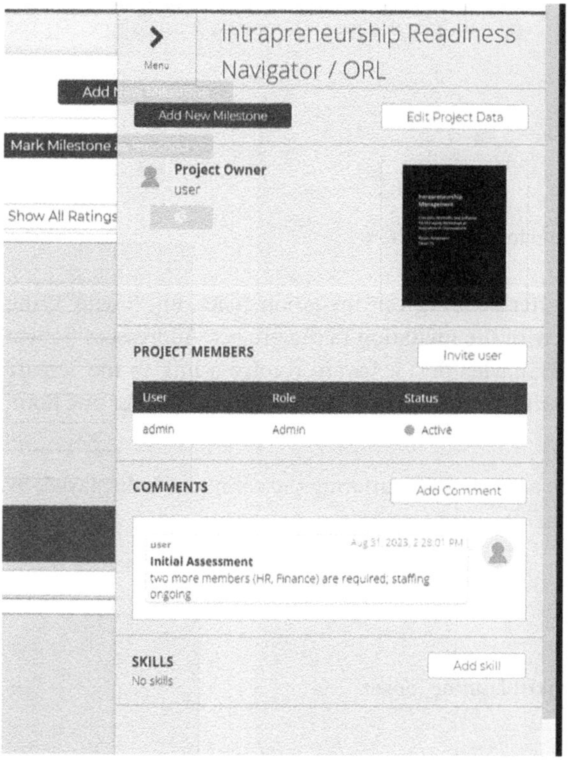

Menu
Intrapreneurship Readiness Navigator / ORL
Add New Milestone
Edit Project Data
Mark Milestone
Show All Ratings
Project Owner
user
PROJECT MEMBERS
Invite user

User	Role	Status
admin	Admin	Active

COMMENTS
Add Comment
user
Aug 31, 2023, 2:28:01 PM
Initial Assessment
two more members (HR, Finance) are required; staffing ongoing
SKILLS
No skills
Add skill

Figure 8.6 (a) Assessment Menu for ORL

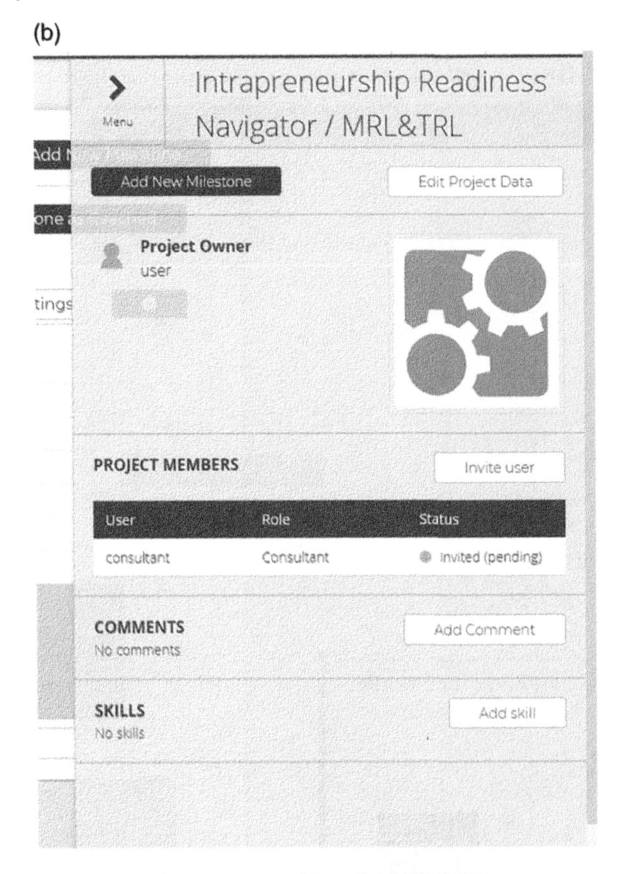

Figure 8.6 (b) Assessment Menu for MRL & TRL

address must be entered. After sending the invitation (click on "invite"), the addressees receive an email with the invitation to the project. Addressees not yet registered in the READINESSnavigator© software receive a link to the "create account" function – the security token contained in this link is valid for one hour.

Comment/Add Comment

After clicking on this button, a window for entering the comment is displayed; by clicking on "leave a comment" the input is saved.

Reference

1 https://en.wikipedia.org/wiki/Planning_poker.

9

Summary and Outlook

This final chapter presents an overall summary of the book and an outlook for the future developments of Intrapreneurship Management.

9.1 Overall Summary

In the 21st century, innovation has been universally recognized as the critical driving force for economic growth, environmental sustainability, as well as social welfare improvements. In this book, we have developed a simple definition of innovation as *a creative idea implemented with significant impact*. This emphasis on implementation directly implies that innovation requires collaboration among Innovator Team, Supporter, and Adopter.

While many innovations have been developed and implemented by innovation teams, or entrepreneurs, that build independent organizations with external support, history has shown most innovations have been developed and implemented by internal innovators, or intrapreneurs, and their internal supporters in existing organizations. However, different entrepreneurs and successful intrapreneurs generally require development and management. Thus, effective Intrapreneurship Management is essential to future human advancements. However, because of the overshadowing glamour of entrepreneurship and the intrinsic complexity of internal innovation, there has been a lack of comprehensive and in-depth studies on Intrapreneurship Management, which has provided the motivation for this book.

Because developing and implementing a creative idea in an organization is by nature a team sport and an iterative process involving Innovation Team, together

Intrapreneurship Management: Concepts, Methods, and Software for Managing Technological Innovation in Organizations, First Edition. Rainer Hasenauer and Oliver Yu.

Published 2024 by John Wiley & Sons, Inc.

with Internal Supporter and Adopter, the book takes an integrated system analysis approach to examine the joint decision process of these major participants based on a common optimal tradeoff between their individual perceived expected values and risks (of failing to achieve the value) for their collective investments in developing and implementing innovation in the organization.

To further understand these values and risks, the book extends Maslow's hierarchy to identify two fundamental dimensions of human needs and wants: physical and psychological needs for survival and security and wants for stimulation and growth. Human values can be classified into four major categories from the basic needs for physical and psychological survival and security to the driving pursuits for physical and psychological stimulations, and the inspiring advancements for psychological growth, with the intensity affected by the availability of economic, intellectual, and belief-based resources. These categorizations can be extended to the collective values of an organization and of a society as a whole. They also form the basis of the Economic–Ecologic–Equity (EEE) integrated framework for innovation value and risk assessment for the participants.

The book then develops a unified approach to marketing by viewing the collaboration among the innovation participants as a mutual marketing process with the common stages of Attracting, Engaging, Convincing, Committing, and Maintaining each other. It then introduces two major tools for Innovation Team building: Agile & Scrum and Enneagram. It further develops the Organization Readiness assessment framework with a list of questionnaires and checklists to assess the readiness level of an organization for effective internal innovation development and implementation.

The book next examines the evolving concepts and practices of marketing to innovation adopters, both individuals and groups. It discusses the importance of marketing ethics for long-term relationships between the innovation participants. It then develops the Market Readiness assessment framework with a list of questionnaires and checklists to assess the readiness level of an organization to successfully market innovation products (goods or services) to a prospective Adopter or customer.

Finally, the book explores the barriers to creative problem solving and examines the root causes for the decrease of creativity in adults and means to reverse the effects including two major proven tools: TRIZ and Enneagram. It then develops the Technology Readiness assessment framework with a list of questionnaires and checklists to assess the readiness level of an organization to successfully develop innovative products (goods or services) to add significant value to the prospective Adopter.

Integrating all these readiness assessment frameworks, the book presents the methodology and the outline of a software for Intrapreneurship Readiness assessment. The software is adaptive and can evolve with additional developments in the various assessment frameworks and user inputs.

9.2 Future Outlook and Potential Research Areas

With increasing awareness and appreciation of the importance of Intrapreneurship by all public and private organizations, there will be growing interest in understanding and instituting effective Intrapreneurship Management concepts and tools. Moreover, developments in understanding of human motivation and analysis of management processes will continue to advance over time. For example, developments in neuroscience have continuously uncovered new bases for the formation of human needs and wants, and rapidly advancing tools such as those provided by artificial intelligence (AI), quantum computing, and nescience are revolutionizing the analytics field. Thus, we expect that there will be many new insights gained and breakthrough methodologies developed in the next decades. In anticipation of these advances, we not only strongly promote continued research on Intrapreneurship Management, but also have deliberately made our methodology and software flexible and easily modifiable to adapt to new and different inputs.

Looking into the future, we have assembled a sample list of interesting and promising research areas for Intrapreneurship Management and Readiness Assessment:

- The first area will be refinements of current methodology, which include:
  - Indicators for readiness of disruptive technologies.
  - Additional research and development on key performance indicators (KPIs) for effective Intrapreneurship Management.
  - Further research and integration of the circular economy readiness.
  - Assessment of Market and Technology Readiness of hard real-time systems which, as opposed to soft real-time systems, deal with system behaviors with serious consequences, such as missing a deadline for a system-critical event, that may cause disastrous, possibly self-destroying effects. Other application examples include control systems of moving objects, such as trains, aircraft, submarines, military objects, orbiting satellites, and space flights, as well as rescue processes under tight time constraints.
- Another area will be the applications of emergent technologies, such as quantum computing, advanced optimization algorithms, new data and communication security, and fake information detection technologies for more accurate readiness assessment.
- Still another exciting area of research is the application of advanced AI as an automated Intrapreneurship Management readiness assessment coach, which both guides the user group through the assessment process with helpful prompts from a large database of past user experiences, and monitors and evaluates the users' intellectual responses and emotional status to elicit additional information, improve communications, build trust, and facilitate productive interactions.

In the meantime, the IRN© software has an open architecture that can not only incorporate any new developments for the assessment of Intrapreneurship Readiness but also serve as a fertile ground for stimulating new concepts and tools for the assessment.

Appendix A

Case Studies of Innovative Companies

This appendix presents case studies of the organizational culture of three well-known and long-lasting innovative companies based on interviews with executives with long tenures and studies of company history literature.

A.1 3M

The following is a summary of interviews with Gerald (Gus) Gaynor, Director of Engineering, 3M Europe (retired after 25 years) and Steven Webster, Vice President, Research and Technology Commercialization at 3M (retired after 32 years), and contents of *A Century of Innovation, the 3M Story* by 3M.

A.1.1 The Company and History of Intrapreneurship Management

3M (originally the Minnesota Mining and Manufacturing Company) is a multinational conglomerate of industrial, worker safety, healthcare, and consumer products. The company was first formed in 1902 to mine corundum as an industrial abrasive but failed because the mine's mineral holdings were anorthosite, which had no commercial value. Fortunately, through the availability of "patient" money, the employee-centric management style of its top executives, quality assurance from direct interactions with the customers, and the continuous technology research, the fledging company not only survived but also started to thrive with many management and technology innovations, highlighted by three representative examples.

Intrapreneurship Management: Concepts, Methods, and Software for Managing Technological Innovation in Organizations, First Edition. Rainer Hasenauer and Oliver Yu.

Published 2024 by John Wiley & Sons, Inc.

A.1.1.1 The 15 Percent Rule

One of the most celebrated management innovations at 3M is the 15% rule that allows employees to spend 15% of their time on projects of their own choosing. The unprecedented policy was instituted in 1948, 56 years prior to Google's 20% rule, by William McKnight who succeeded the founders as President of 3M from 1929 to 1949, then Chairman of the Board from 1949 to 1966, and Honorary Chairman until 1972. McKnight was well-known for encouraging 3M management to delegate responsibility and encourage employees to exercise their initiative, and his management theories are still the guiding principles for 3M. His basic rule of management was laid out in a 1948 interview with New York Times:

> As our business grows, it becomes increasingly necessary to delegate responsibility and to encourage men and women to exercise their initiative. This requires considerable tolerance. Those men and women, to whom we delegate authority and responsibility, if they are good people, are going to want to do their jobs in their own way. Mistakes will be made. But if a person is essentially right, the mistakes he or she makes are not as serious in the long run as the mistakes management will make if it undertakes to tell those in authority exactly how they must do their jobs. Management that is destructively critical when mistakes are made kills initiative. And it's essential that we have many people with initiative if we are to continue to grow.

Here, McKnight envisioned a "flat" organization decades before the concept was a popular business model. His philosophy led to tenets of the 3M culture: minimal hierarchy, intentional informality, and strong support for creativity and innovation.

By allowing people to make the right decisions on their own and being rewarded for taking initiative, the 15% rule has induced extraordinary mutual respect and trust between management and employees at 3M on their respective integrity and caring at an age when command-and-control management style was the norm. Such mutual respect and trust in turn generates strong company loyalty and internal collaboration to promote corporate success and common good, which was fully echoed by the two 3M executives interviewed for the case study. It provides a major foundation for successful Intrapreneurship Management at 3M.

The development and implementation of Post-It notes exemplifies such loyalty and collaboration. The following is excerpted from *100 Years of Innovation, the 3M Story:*

> In 1960s, Dr. Spencer Silver, a 3M senior scientist, was studying adhesives and discovered an adhesive that didn't act like any others. Instead of

forming a film, this adhesive turned into clear spheres that, according to Silver, "kind of sparkled in the light." Silver spent the next few years shopping his new adhesive around 3M to find a product use for it, but the reception wasn't stellar. In other companies, this might have been discouraging enough to scrap the idea, but Silver didn't give up.

Five years after Silver's initial discovery, Art Fry, another 3M scientist was warming his vocal cords while sitting in the choir loft at his church. Frustration rose with his scales as Fry turned to a hymn and his scrap paper bookmark fell to the floor. "My mind began to wander during the sermon," Fry confessed. "I thought about Spence's adhesive. If I could coat it on paper, that would be just the ticket for a better bookmark."

Fry went to work the next day, ordered a sample of the adhesive and began coating it on paper. He only coated the edge of the paper so the part protruding from his hymnal wouldn't be sticky. "When I used these 'bookmarks' to write messages to my boss, I came across the heart of the idea. It wasn't a bookmark at all, but a note," said Fry. "Spence's adhesive was most useful for making paper adhere to paper and a whole lot of other surfaces. Yet, it wasn't so sticky that it would damage those surfaces when it was pulled off. This was the insight. It was a whole new concept in pressure-sensitive adhesives. It was like moving from the outer ring of the target to the bull's eye."

Fry encountered serious technical problems very early and his boss, Bob Molenda, encouraged him to takes things one step at a time. First, there was the problem of getting the adhesive to stay in place on the note instead of transferring to other surfaces. And, although 3M was known for its coating expertise, the company did not have coating equipment that could be precise on an imprecise backing such as paper. It was difficult to maintain a consistent range of adhesion. "All of these things bothered our production people," Fry said, "but I was delighted by the problems. If there is anything that 3M loves, it's to create a product that is easy for the customer to use but hard for competitors to make." Fry used his 15% time to find manufacturing and technical solutions over about 18 months, and Molenda helped Fry find the time and money to dedicate to his pet project.

It would be years before the Post-it notes adhesive was perfected, prototypes created, and the manufacturing process developed. All the while, Fry busily handed out product samples and Geoff Nicholson, then Fry's technical director, made sure that secretaries of 3M senior executives got them. Before long, their bosses were borrowing the little yellow pads. Everyone who tried them wanted more. In 1977, with a host of product literature in tow, 3M conducted market tests in four major cities. But, consumers were lukewarm at best. Ramey was a new division vice president when Post-it notes tested so poorly. Seeing how many 3Mers truly believed in the product, Ramey decided to figure out why the notes weren't faring well. He and Nicholson traveled to a test market and met with people, only to discover that advertising and brochures weren't good enough. What consumers really wanted was the product itself. Sampling, however, was an expensive proposition – especially for a product with a questionable future. Ramey bypassed the traditional approval channels and went straight to Chairman of the Board and CEO Lew Lehr to fund the Post-it Note sampling.

In 1978, 3Mers descended on Boise, Idaho, with samples for what would later be called the "Boise Blitz." The town was a perfect venue – not too big, not too small, and remote enough to truly be able to measure results accurately. Sample upon sample were handed out, and 3M discovered that more than 90% of the people who tried them would buy them. With success in Boise, 3M was convinced that the market potential for the yellow note was enormous, and, in 1980, Post-it notes were introduced nationally.

For their efforts, the Post-it Note team was awarded the Golden Step Award, the highly coveted internal award recognizing teams that develop significant profitable products generating major new sales for 3M. By meeting this criteria twice, the team won the award two years in a row, 1981 and 1982.

This innovation development and implementation history vividly illustrates how the 15% rule and the resulting mutual respect and trust between management and employees have not only provided the needed support and resources to the internal innovators, but also promoted the close collaborations between Silver and Fry and among the research, production, and marketing departments at 3M to achieve the significant impact for this innovative product.

A.1.1.2 The Tech Forum

Another major 3M management innovation was the Tech Forum. The Forum was founded in 1951 to promote 3M technical employees in different fields worldwide to collaborate, educate, and learn from each other. Tech Forum is a grassroots organization led by the Tech Forum Senate with representatives from each division, business group, and lab group within 3M and supported by special-interest chapters related to the company's technology platforms. It focuses on providing education to the 3M technical community on a broad range of topics, from internal expertise, and scientific curiosities, to rising market demands. It also enables its members to bring in external experts from academia, government, and other entities through funding and executive support.

Through this system, Tech Forum empowers 3M technical staff members to harness the synergy of creative thinking and stimulating interactions that drive the innovations at 3M through a comprehensive network of technical events and symposia, recognition programs including the prestigious Carlton Society and Young Inventors Recognition, and other resources like the 3M Genesis Grants and Discover Program for projects not readily funded through normal channels or outside the researcher's primary responsibilities.

The synergy and collaboration created by 3M Tech Forum are further enhanced by the 3M Technical Skills Database, which allows members of the 3M technical community to find a 3M expert around the world associated with 3M's key technologies so that they can answer technology or research-related questions and share knowledge on a challenging program or new idea. This proprietary database contains expertise profiles for nearly 10,000 Tech Forum members and highlights their patents, invention submissions, and approved publications.

The most important impact of the Tech Forum is the inspiration and motivation provided to the employees. As related by one of our interviewees, Steven Webster, when he joined 3M soon after college, he was greatly inspired by the exciting technical events and motivated by recognition programs to want to "personally make a difference for 3M."

A.1.1.3 Social Equity and Responsibility Programs

In addition to the 15% rule, 3M has been a leading innovator in many social equity and responsibility programs. The following are a few historical examples:

1916: Employee profit sharing plan

1930: Employee pension plan

1950: Employee stock purchase plan

1953: Corporate foundation for community giving and social responsibility

1960: Hiring the first PhD employee, Dr. Joseph Ling, a Chinese-American, to environmental programs that led to the Pollution Prevention Pays program with the innovative idea of preventing pollution in 3M products and processes rather than cleaning it up later

1972: Employee assistance program

1977: One of the first American companies to adopt the Sullivan Principles, a voluntary code in which the companies pledged to practice fair and equal employment practices as well as contribute to the progress of black South Africans in the workplace and community. 3M South Africa was one of the first companies to integrate its workplace, even though local laws required segregation; and publicly pledged to "play an active role" in ending apartheid. The innovation here was, instead of pulling out of South Africa like many other US companies, 3M chose to stay to be a positive force for change that eventually proved to be more effective than disengagement

1980: Employee diversity program for women and minorities

In summary, with its extraordinarily innovative culture, 3M has long reached Organization Readiness and has been a model for effective Intrapreneurship Management.

A.1.2 Insights from the Interview

Steven Webster started at 3M as a technical staff member and was gradually promoted through the management track. Based on his long years of experience, both as an Intrapreneur and an Internal Supporter, he offered the following insights about motivating and managing internal innovation and becoming a successful Intrapreneur at 3M.

Steve listed the major factors that attract and motivate Intrapreneurs at 3M as:

- The Tech Forum which is organized at the local chapter level and provides exciting opportunities for learning and collaboration in the technical community.
- The 15% rule gives Intrapreneurs the freedom to explore innovations of their own interest.
- The recognition and celebration of the top innovators, or heroes, at 3M, which often inspire Intrapreneurs on the personal mission to "do something for the company or the world."

To be successful, Steve indicated that an Intrapreneur needs to build broad alliances with not only technical collaborators but also business professionals, especially the marketing staff, to gain understanding of customer needs. The Intrapreneur then needs to develop a clear vision that can generate excitement as well as the political savvy and tenacity for working with the system to form consensus to gain support from senior management.

On the other hand, a major challenge for the Internal Supporter has been how to assess the credibility and probability of success of an innovative idea, as there are no reliable key performance indicators. However, it usually helps to have the idea presented to and critiqued by a broad audience with different perspectives.

One other insight Steven provided is that, as an internal innovation project becomes successful, the original Intrapreneur who started the project may not be the best match for the project. In this case, the management as an Internal Supporter needs not only to recognize the change but also devise innovative ways to channel the energy and talent of the original Intrapreneur to more productive activities, such as another new creative idea or innovation project.

A.2 IBM

The following is a summary of information gleaned from interviews with Dr. Jim Spohrer, Director, IBM Cognitive Opentech Group (retired after 22 years) and Dr. Zong Ling, Senior Research Scientist at IBM (retired after 20 years), and the contents of IBM 100, a website celebrating 100 years of innovation at IBM.

A.2.1 The Company and History of Intrapreneurship Management

IBM (International Business Machines Corporation) is an over a century-old multinational technology corporation with about 300,000 employees and over $60 billion in gross revenues worldwide as of 2022. Like 3M and Google, the initial leaders of IBM in its founding and formative years, Thomas Watson, Sr. and his son Thomas Watson, Jr., CEOs in 1924–1956 and 1956–1971, respectively, have fostered many management innovations to develop a strong innovative organizational culture. The following are a few representative examples:

A.2.1.1 A Professional Sales Force

Recognizing the important advantage of a professional sales force that knows not only about the product but also about the company and the industry, Watson Sr. hired top graduates from the best universities as professional salespersons. He set up a sales school, putting the trainees through six weeks of intensive training in selling and servicing IBM equipment. He insisted that IBM salesmen wear conservative clothing and conduct business to the highest ethical standards. He also ensured diversity of the salesforce – women, black, disabled, etc.

A.2.1.2 Extraordinary Emphasis on Research

By hiring engineer and inventor James W. Bryce in 1917, Thomas Watson Sr. showed his commitment to pure inventing. Bryce and his team established IBM as a long-term leader in the development and protection of intellectual property. By 1929, 90% of IBM's products were the result of Watson's investments in R&D. To further the basic corporate belief in excellence, Watson Jr. put extraordinary emphasis on innovation by significantly expanding research centers and promoting patent applications. In 2010, IBM ranked number one among companies receiving US patents, with 5,896 granted. It marked the company's 18th consecutive year as the US patent leader.

A.2.1.3 Strict Adherence to Business Ethics

As a sales professional who long recognized the importance of trust by the customers as well as the temptation for salespeople to exaggerate or bypass the truth, Watson Sr. instilled in IBM employees a strict adherence to business ethics which has continued over the years and has become a hallmark for IBM around the world. In fact, a retired 3M executive, Steven Webster, recalled how IBM would meticulously avoid any possible conflict of interest in a collaborative effort between the two companies. Moreover, one of the interviewees of this case study, Dr. Zong Ling, a retired IBM senior scientist, was invited by the Chinese government to develop a course on business ethics to be used for major state-owned enterprises. IBM requires all employees each year to review and sign business conduct guidelines.

A.2.1.4 Pioneer in Social Equity and Corporate Responsibility

Watson Sr. has long believed in social equity and corporate responsibility. The following are a few historical examples of the pioneering programs at IBM:

1914: First disabled employee
1916: Employee education
1933: 40-hour week
1934: Elimination of piecework
1936: First professional saleswoman
1937: Paid holidays and vacation
1943: First female vice president
1946: First black salesman
1962: Company-wide focus on environmental programs

In summary, IBM has achieved Organization Readiness in Intrapreneurship Management from the beginning due to its dedication to basic values or core beliefs of excellence, ethics, and equity, which, in turn, has attracted talented Intrapreneurs to transform it from a small regional business to an international industrial giant.

A.2.2 Insights from the Interview

Dr. James Spohrer has been a highly successful Intrapreneur at IBM, who has been cited in IBM 100 for creating service science. After an initial stint as a distinguished engineer at Apple, Jim joined IBM in 1998, being attracted, like many other potential Intrapreneurs, by IBM's extensive reputation and successes in innovation, ethics, and equity, and motivated by IBM's well-established career path, ample financial resources, and vast technical network, and even more importantly, "the opportunity to change the world."

Jim attributed his success at IBM to the valuable advice given to him, when he first joined IBM Almaden Research Center in San Jose, California, by its director, Dr. Paul Horn, which was to "leverage the matrix at IBM."

Based on Jim's experience, there are several steps to achieve this leverage:

First, align personal values with company values. An established company like IBM generally has well-developed corporate values, internal culture, and organizational structure. An Intrapreneur must align these with their own values to work and interact effectively.

Second, build a reputation of expertise and capability by achieving through the technical career ladder of a senior member of technical staff, distinguished staff member, and IBM fellow. The latter two levels have executive powers that engender greater responsibilities and influences in helping IBM secure clients and sales, cultivating young talents, formulating business strategies, and steering technical developments to success.

Third, build alliances with not only other technical staff but also business staff. This diverse collaboration will not only broaden the Intrapreneur's perspective but also deepen the understanding of customer needs and business requirements. Business staff includes sales, legal, finance, human resources, procurement, supply chain, and more – learn the whole business and have good contacts in every part of the business.

Finally, support executives first to receive support in return. Specifically, ask executives about projects they need to get done and help get one or more of them done. This step demonstrates the Intrapreneur's value to the executive as well as the EQ for working in an established organization, which is crucial to the successful mutual marketing between Intrapreneur and Internal Supporter.

Regarding the potential complacency or bureaucracy of a large organization like IBM, Jim believes that IBM's large resource base can endure and survive missteps or setbacks much better than a small startup. Furthermore, with IBM's vast talent base, it can continue to thrive if a savvy CEO knows how to strategically sell off declining businesses and build or buy rising business units.

A.3 SRI International

The following is based on a summary of interviews with Dr. Curtis Carlson (retired after 28 years with the last 16 years as President and CEO) and Michael Gold (retired after 30 years as Senior Research Engineer), and personal experience of Dr. Oliver Yu (retired after 21 years as Director of Energy and Technology Strategies).

A.3.1 The Company and History of Intrapreneurship Management

SRI International was founded by Stanford University in 1946 as Stanford Research Institute with the purpose of providing a research facility for its faculty members for the benefit of the public. To generate funding at a time when US defense research was rapidly expanding during the Cold War years, the institute rose quickly to become a major independent contract research organization for US government agencies, especially the Department of Defense. Since then, more than 80% of SRI's annual revenues have come from government-sponsored research projects. In the meantime, Weldon Gibson, a Stanford Ph.D. and the third employee of the institute leveraged the recognition and resources of Stanford University to successfully develop business consulting projects around the world to build a strong international reputation for the institute. With defense research funding, especially from the Defense Advanced Research Project Agency (DARPA), SRI has developed many breakthrough innovations, including ARPA-net, the initial version of the internet; computer mouse; Window technology; and speech recognition and robotic operation technologies. For business innovations, SRI has developed the Values and Lifestyle Survey (VALS), a marketing tool based on the psychographic segmentation of customers, and scenario planning technique.

During the Vietnam War years, Stanford students strongly protested against connections between the university and any military research. As a result, Stanford University severed relations with the institute, and the latter was renamed SRI International, but required to pay 2% of its annual gross revenues as a tribute to the university for its founding. This requirement was terminated only in the late 1990s when SRI was facing insolvency as defense research had waned.

As a contract research organization founded by an educational institution, the initial innovation management style at SRI had been informal and opportunistic. For decades, it often acted as a consortium of independent contract researchers supported by a common administrative structure. The nominal 3% of government contracts earmarked for the institute's R&D fund were usually not sufficient to support serious internal innovations. On the other hand, successful research programs, like advanced computer technologies, power system control software, and decision analysis methodology, would spin off without compensation to SRI and sometimes even become competitors to SRI. Consequently, when defense research funding declined in the 1990s, SRI encountered severe financial difficulties. In 1998, the organization was on the verge of bankruptcy when Dr. Curtis Carlson took over as CEO.

A.3.2 Insights from the Interview

Dr. Curtis Carlson has been a consummate Intrapreneur. After receiving his Ph.D. degree from Rutger University in 1973, he joined Sarnoff Laboratory, the corporate research arm of RCA. In 1988, General Electric acquired RCA and subsequently donated the laboratory, which was renamed Sarnoff Corporation, to SRI International as a wholly-owned subsidiary. At the time of the transition, Curt was already the Director of the Image Quality and Perception Research Group. Afterward, he rose to Vice President in 1990, and Executive Vice President of the

Interactive Systems Division at Sarnoff. He started the 1997 team that developed the HDTV program that became the US standard. He also started the 2000 team that designed a system to assess broadcast image quality. Both of these teams were awarded a Technology & Engineering Emmy Award for their accomplishments. In 1998, Curt became the longest-serving President and CEO at SRI. During his 16-year tenure, Curt transformed SRI by tripling its revenues to become very profitable and created and spun off many world-changing innovations, including Siri and Intuitive Surgical. He achieved these through the approach that he summarized in *Innovation: the five disciplines for creating what customers want*, a best-selling business book he co-authored with William Wilmott in 2006. Since retirement in 2014, Curt has continued as Founder and CEO of Practice of Innovation, LLC, a company working with start-ups, established companies, and government agencies on improving innovative performance.

The following is a summary of the interview with Curt about the development and implementation of his proven process for effective innovation and insights gained on its benefits.

A.3.3 Development History

When Sarnoff Lab became a subsidiary of SRI, Curt realized that instead of conducting corporate-funded research, he would need to seek funding from external customers. He and his team of 15 staff members decided to *learn* how to get funded by providing value to potential customers. Through many Monday night meetings, they finally developed a Value Creation Forum (VCF) process for innovation that he has since successfully applied and perfected over the years, first at Sarnoff, then at SRI, and now at Practice of Innovation. The first application was for developing a highly successful analog-to-digital transfer technology for the intelligence community. Afterward, he continued at Sarnoff because of the freedom and independence and the talented colleagues at Sarnoff, which he indicated as important values for attracting and retaining Intrapreneurs.

A.3.4 The Value Creation Forum Process and its Implementation

The VCF process requires the senior management and the team involved in a project to continuously refine the fundamentals, i.e., the need, approach, benefit/cost, and competition (NABC) elements of the project through 10-minute presentations at regularly scheduled VCFs. After each presentation, the VCF facilitator would randomly ask for feedback from other participants in the process to improve the presentation.

During his tenure as president and CEO of SRI, all employees were required to attend a three-day workshop to learn the VCF process. Then, for about five or six major prospective or existing projects from each division, the senior management and the team involved in each project were required to present and refine the NABC of the project at a series of biweekly VCFs.

A.3.5 Benefits of the VCF Process

Both a professional and a company must continually create value for the customer to survive and thrive. Innovation is the process of creating value.

Unfortunately, most CEOs are managers, not innovators. Without continuous innovation, a company merely lives off the benefits of past innovations. That is why an average company only lasts for about a decade.

The process provides a superb platform for engagement, collaboration, communication, and transparency by providing the participants with shared language and concepts. It also strongly motivates and empowers Intrapreneurs and enables the Internal Supporters through interactions.

The process is also a powerful training tool for clear thinking and effective communication.

For motivation, the process should start with *important* customer needs. A customer may sense a problem but not necessarily know the need, because it does not know what is possible.

Two effective ways to determine customer needs: one is to closely observe, interact, and build empathy with the customer, and the other is to conduct iterative *reframing* from different perspectives on the customer through repeated presentations at the Forum.

The process also continually improves the innovative solution approach of the project through continuous feedback from VCF.

The process allows the team time to improve their understanding of customer needs and search for additional solution approaches between presentations.

With the transparency provided by the process, things that do not make sense will go away, interpersonal issues will be cleared, and organizational problems will be resolved.

An effective team requires at least a member with business acumen and another member with technical expertise, with the support of a member with operational capability. The process can also identify the strengths and deficiencies of the team makeup.

A champion is critical to the project, and the enthusiasm and determination can also be assessed through the process.

An implementer of the VCF process for an organization should receive royalty for its contribution to the efficiency and profit gained from the resulting productivity improvements.

Finally, understanding and instituting the VCF process by top management can be very helpful in stimulating innovation in an organization from the top.

However, based on Curt's successful experience, an aspiring Intrapreneur can be equally effective by starting the process from the bottom through a grassroots effort with a small team to achieve valuable innovations and then spread the process throughout the organization to change culture and scale success.

A.4 Summary of Insights

Through extensive studies of the three innovative companies and interviews with their senior executives, we have gained the following insights on internal innovation in an organization.

A.4.1 Innovation is the Life Force for Sustained Growth of an Organization

All these three companies have long surpassed the average 15-year lifespan of major companies in the United States. As pointed out by Dr. Curtis Carlson, former CEO of SRI International, "Without continued innovation, a company merely live off the fruits of past innovations and will eventually die off." Therefore, a company must develop a truly innovative culture, not just lip service, to incentivize internal innovation in order to survive and thrive.

A.4.2 Innovative Culture Stimulates Internal Collaboration

One of the most important values of an innovative culture is its ability to build mutual trust, pride, and camaraderie among members of an organization. In such a culture, people feel they are respected and also energized by a common vision of providing value to not only the Adopter of innovative products but also the society as a whole. The extraordinary collaboration among technical and business staff members in the development and implementation of Post-It innovation is a typical successful example of the impact of an innovative culture. On the other hand, without an active innovative culture, the aspiring internal innovator will often feel discouraged and disappointed, and eventually leave for more fertile ground to the detriment of the company.

A.4.3 Internal Innovators Need to Market Themselves

Even for a company with strong innovative culture and well-developed support system, an aspiring internal innovator or Intrapreneur, still needs to do internal marketing to attract attention and earn trust and, more importantly, to provide value to Internal Supporters to win their support. Although Internal Supporters should proactively market the company and themselves to potential internal innovators, they often have many competing priorities as well as certain risk adversity. Innovation is about providing value. Therefore, to be successful, internal innovators must first market themselves to Internal Supporters to start the two-way street.

A.4.4 An Intrapreneur can be Successful by Creating Internal and External Values

Both 3M and IBM have had long histories and well-established corporate policies for building innovative culture and providing support to aspiring internal innovators. On the other hand, as a contract research organization, SRI International is much less organized and passive in developing Intrapreneurs, although, through its research focus, it does encourage free thinking, creative pursuit, and innovation collaboration, as long as a staff member can find an Internal Supporter. It is thus remarkable to see that an Intrapreneur like Dr. Curtis Carlson can be successful and rise to the top to change the culture and expand the company by learning how to effectively create value for both internal and external customers.

Appendix B

Selected Cases for Innovation Project Readiness Assessment

The Market- and Technology Readiness assessment methodology presented in Chapter 7 has been implemented as a part of the READINESS navigator (RN©) software since 2010 and applied to over 200 innovation projects. In this appendix, we will present five diverse selected cases from a collection of 26 innovation projects to demonstrate the effectiveness of the methodology in guiding the development of innovation projects.

B.1 General Overview

This appendix first presents an overview of the collection of 26 innovation projects and the general results of applying the methodology in Chapter 7 through the predecessor of the Intrapreneurship READINESS navigator (IRN©) software to assess the Market Readiness level (MRL) and Technology Readiness level (TRL) of these projects over a span of about 20 years.

Because the assessments have been conducted for the projects at various stages of their development, we will present a collage of the Market and Technology Readiness assessment results of the 26 innovation projects collected at different times to provide a general overview of how the assessment may help the development of innovation projects. We will then present the detailed descriptions and the evolutions of the assessment results for five selected cases from this collection at the start, midpoint, and finish of each selected project to show how the assessments actively guide the development of these projects.

Intrapreneurship Management: Concepts, Methods, and Software for Managing Technological Innovation in Organizations, First Edition. Rainer Hasenauer and Oliver Yu.

Published 2024 by John Wiley & Sons, Inc.

B.2 Summary of the Collection of 26 Innovation Projects

The collection of 26 innovation projects for different industries is shown in Table B.1, with the selected project shown in ***bold italics***.

Representative market and technology assessment results for the 26 innovation projects are summarized in Table B.2. These results are taken at various stages of the development of these projects to show how the assessment can assist the project development at various stages. For Market Readiness, we have added a level 10 indicating the product has been well-accepted by the market and the marketing strategy is ready to be integrated with the business model.

Table B.1 Summary of the Collection of 26 Innovation Projects

ID	Innovation	Industry	ID	Innovation	Industry
A	Gesture Controlled MMI	Scanner	N	Continuous blood pressure measurement	Medical diagnosis
B	Technical Simulation	Software	O	"Watchdog" for semiconductor	Software
C	Atmospheric Nitrogen Deposition Collector	Sensor	P	Containment	Building construction
D	Aerosol Jet-Printing	3D printing	Q	Wearable energy meter	Sensor
E	Selective Laser Melting	3D printing	R	Lab-on-chip diagnostics	Software
F	Sensors for Mobile Robots	Sensor	S	Vibrational acoustic analysis	Medical diagnosis
G	Health CCPM and Industrial ACT	Robotics	T	Smart bottling plant	Machine construction
H	Safety Robot	Robotics	U	Bright red systems	Scanner
I	Atmospheric Plasma on Wood Surface	Material science	V	Pressure and temperature sensor	Sensor
J	Phase Change Material	Building construction	W	Bionic surface	Material science
K	Flame Retardant Rubber	Material science	X	Cellular materials	Material science
L	Magic Lens Augmented Reality	Software	Y	Vanadium–redox flow battery	Energy storage
M	Bone Diagnostics	Medical diagnosis	Z	Diamond-like carbon	Material science

Table B.2 Market and Technology Readiness Assessment Results of 26 Innovation Projects at Various Stages of Development

Readiness of 26 Technology Push Projects

Market-Readiness	Demand Level	1	2	3	4	5	6	7	8	9
Building the adapted answer to the expressed need in the market	9						*L*		H	
Identification of the Experts possessing the competencies	8					*Q,U*	*B*		JK	Y
Definition of the necessary and sufficient competencies and resources	7									
Translation of the expected functionalities into needed capabilities to build the response	6					O	M,T	*P*		
Identification of system capabilities	5				Z	A				
Quantification of expected functionalities	4		W		F,V	R	N	*E*		
Identification of the expected functionalities for new product/service	3			S	X	C	G	D		
Identification of specific need	2		I							
Occurence of feeling "something is missing"	1									
	Technology Level	Fundamental research	Applied Research	Research to prove feasilibility	Laboratory Demon-stration	Technology Development	whole system Field demon-stration	Industrial Prototype	Product Industrial-isation	Market/Sales Certification

Technology Risk

Market Risk

Industry	
Software	BLOR
Sensor:	CFW
Material	IKWXZ
med. diagnosis	MNS
Scanner	AU
Robotics	GH
3D print	DE
building constr.	JP
wearables, location services	QY
machine	T

Italic not ready for market
GREY- Transition:
Black-Ready for market
Off diagonal = risk
Technology management = stay on diagonal!

Insights of Readiness Assessment to Innovation Project Development:

Northwest quadrant: Technology not ready and potential intense competition.
Southeast quadrant: Market not ready and potential "white elephant" should be abandoned.
Northeast quadrant: Market and technology ready and potential "sweet spot".
Southwest quadrant: Market and technology under development and potential watchlist.

Innovation management principle: Stay on the diagonal.

Specifically, projects with MRL/TRL in the northeast quadrant are in the "sweet spot" or "window of opportunity" region discussed in Chapter 5. They are ready for continued technological refinements and business model development ready for the market. On the other hand, projects in the southwest quadrant are exploratory projects that need further market research and technology development to move into the "sweet spot." Some of these projects may be put into the monitoring status as they are still far away from being economically practical.

As discussed in Chapter 5, projects in the northwest quadrant have strong market demand but also large technology risks for not being able to meet the market need. Extraordinary efforts will be needed to successfully beat strong competition. Finally, projects in the southeast quadrant will have high market risks, as they are often solutions in search of problems and are very likely to become "white elephants" with no practical values. They can also sometimes be vanity projects of either the Intrapreneur or the Internal Supporter, which would be projects of strong personal interest but with little consideration for the needs of the Adopter or the organization. In these cases, the senior management of the organization needs to have both the wisdom and strength to terminate these projects.

B.3 Summary of the Five Selected Cases and Readiness Assessment Results

Five cases have been selected from the collection of 26 projects with project IDs: G, H, I, M, and Y for in-depth analysis. They were selected based on their diversity and representativeness. For these five selected cases, their respective start, midpoint, and final MRLs and TRLs from readiness assessments are summarized in Table B.3 and graphically presented in Table B.4.

Table B.3 Summary of Initial and Final MRL and TRL of Five Selected Innovation Projects

Case	Innovation Project	MRL Start	TRL Start	MRLMid	TRL Mid	MRL Final	TRL Final
1a	CCPM robot	1	2	2	4	5	7
1b	ACT robot	3	5	5	7	10	9
2	Safety/security robot	2	2	5	4	8	9
3	Atmospheric plasma on wood surface	1	2	2	3	2	3
4	Bone density warning	2	3	6	6	10	9
5	Vanadium–redox flow battery	2	1	3	5	6	8

Table B.4 Market and Technology Readiness Assessment Results of Five Selected Cases

Integration of Innovation Marketing Strategy with Business Model	10									1bF 4F
Response to the Expressed Need in the Market	9									
Identification of the Experts Possessing the Competencies	8									2F
Definition of the Necessary and Sufficient Competencies and Resources	7									
Translation of the Expected Functionalities into Needed Capabilities to Build the Response	6						4M		5F	
Identification of System Capabilities	5				2M			1aF 1bM		
Quantification of Expected Functionalities	4									
Identification of the Functionalities for New Product (Good/Service)	3					1bS 5M				
Identification of Specific Need	2	5S	2S	3M 3F	1aM 4S					
Unsatisfied Needs have been Identified	1		1aS 3S							
		1	2	3	4	5	6	7	8	9
	Technology Readiness Level	Fundamental research	Applied research	Research to prove feasibility	Laboratory demonstration	Technology development	Whole system field demonstration	Industrial prototype	Product industrialization	Market/sales certification

S, Start; M, mid; F, final.

B.4 Highlights of the Selected Innovation Projects

Highlights of these five selected innovation projects are summarized below, and details of the evolution of the project development and assessment will be presented in the next section.

Case 1 Continuous Compliant Passive Motion (CCPM) Robot and Active Compliant Technology (ACT) Robot

This robotic solution was originally developed for CCPM to support pain-free movements after shoulder surgical intervention. It was accepted in the market as a reliable prototype. However, the internal innovator in the company learned from industry that sensitive robot movements with active compliant technology (ACT) to meet industrial requirements presented a much larger market. Consequently, through the readiness assessment, the Intrapreneur changed focus to industrial solution requirements in sensitive and high precision surface grinding and finishing of surfaces in automotive, aircraft, and other sensitive surface industries. Today, for these new markets, the MRL has reached 10 and the TRL has reached 9. As a result, the abrupt change in market focus and technology trajectory in 2016 proved to be the right strategic decision.

Case 2 Safety Inspection Robot

The versatility of the application fields of this inspection and manipulating robot in harsh, toxic, and explosive environments derives from the deep engineering understanding and systematic reliable design to prevent the impact of harsh, chemically aggressive, or explosive environments. This robot was the first intrinsically safe and the EU "ATmosphere EXplosible" ATEX-certified robot, fully compliant to work in explosion-prone atmospheres like in coal mining, oil refineries, or chemical plants with potential leakages of chemicals, and toxic or explosive gases. The readiness assessment has helped the company to focus on its technology strength in meeting a critical market need. The company is now a partner of a world-leading service company in the oil and gas industry.

Case 3 Atmospheric Plasma on Wood Surface

This innovation project is in the stage of early exploration, close to basic research with potentially disruptive characteristics. The technology is to substitute chemical adhesive energy by activating adhesive energy of surface molecules using high voltage (26–50 KV) and short exposure time of two to three minutes without further preparation of the planed wood surfaces. From an environmental and circular economy viewpoint, such substation is revolutionary. The assessment of the project started with MRL = 1 and TRL = 2, and the market was uncertain, and technology was still being explored. Due to the large number of wood surfaces, extensive experiments were prohibitively expensive. Hence, the research was postponed when MRL reached 2 and TRL reached 3 after considerable efforts. Yet, because of its revolutionary potential, the technology is currently on the continued watchlist.

Case 4 Image Biopsy Lab

This case provides an example of how nescience, or the lack of knowledge, can be transformed into knowledge in the field of finding early warning signals for structural bone disease by examining the X-ray of a tibia bone, which has long been a strong market need in the medical field. Improving the efficacy of using computers for digital examination of X-rays has been a breakthrough and has greatly increased the Technology Readiness of the project to meet the strong Market Readiness. However, unforeseen medical regulations still pose an obstacle to full Market Readiness.

Case 5 Flow Battery with Liquid Electrolyte

Low energy density, low self-discharge, high stability, high reliability, and longevity have made electrolytes a potentially attractive intermediate energy storage for Remote Area Power Supply (RAPS). The project has been under continuing readiness assessment for more than five years in the development of the whole system, including hardware, software, fluid management, electronics and electrochemical design, and functional material design. Long-term functional tests under harsh environments and real-time traffic surveillance conditions showed high reliability and minimum maintenance requirements. Photovoltaic and wind power energy input and a 24-month test operation under demanding temperature and air humidity climate down to −22 °C at night and wind speed up to 120 km/h challenged the steady-state operation. The long operational test was successful with promising Market and Technology Readiness and the whole system and full documentation of know-how was sold to a European system producer.

B.5 Details of Selected Cases of Innovation Project Readiness Assessment

B.5.1 General Approach

For all selected cases, consistent with the discussions in Chapters 5 and 7, a market test bed (MTB) was set up to develop marketability criteria and test the adaptability of the innovation project. It then proceeded to assess iteratively the Market and Technology Readiness of the project.

B.5.2 Case Details

Case 1a Continuous Compliant Passive Motion (CCPM) Robotics by FerRobotics

Company: FerRobotics Compliant Robot Technology GmbH

FerRobotics, headquartered in Linz, Austria, is a world leading developer of lightweight robots. It has developed strong robotic technology capabilities and experiences since 2006. A major capability of the company is in the development of CCPM lightweight robot arm to train people to proactively avoid pain points by sensitively adapting robot motion to the decelerated movements of a patient expecting to touch the individual pain region.

FerRobotics was looking to expand the market for this leading technology capability. Dr. Hasenauer, a co-author of this book, was brought in to help first develop a MTB of potential markets.

The initial prospective market was motivated by the awareness of the strong need due to shortages of skilled and resilient medical staff for physiotherapy to train patients at homecare to regain their physical motility with their shoulder joints after surgical intervention. Because it was a new application market, the starting MRL was one and the starting TRL was 2 due to the inherent technological strength of the company.

Through MTB, market research based on readiness assessment showed that contact sensitive and compliant robotic movements to shape and finish high quality surfaces were potentially a much larger market. This was followed by analysis of competitors and technology acceptance. The first clinical study was successful. Subsequently, the huge market for sensitive feeling robots was opened for FerRobotics after the development of ACT being protected by patent which, in turn, provides legal protection of the following unique selling proposition (USP):

- High physical compliance of the robot, real-time tactile robotic "feeling"
- Magic muscle control
- Based on the patented ACT work along the bionic principle of flexibility and adaptability
- In the therapeutical context, pain sensations are almost certainly avoided

With successful technology developments and additional market understanding, MRL was increased from 1 to 2, and then to 5, and TRL went up from 2 to 4, and then to 7 (Figure B.1).

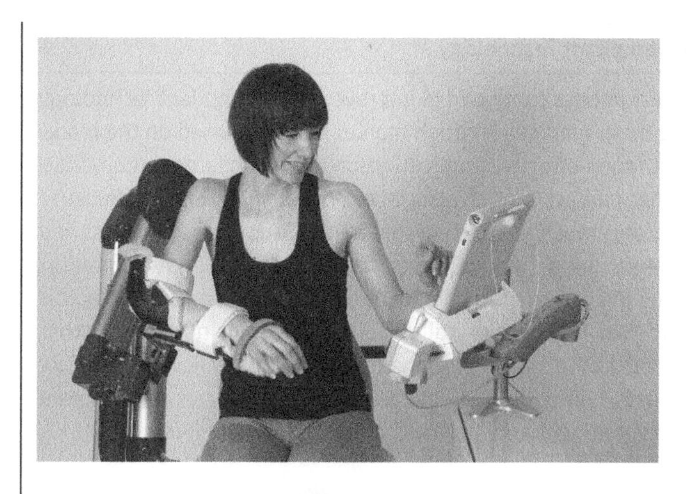

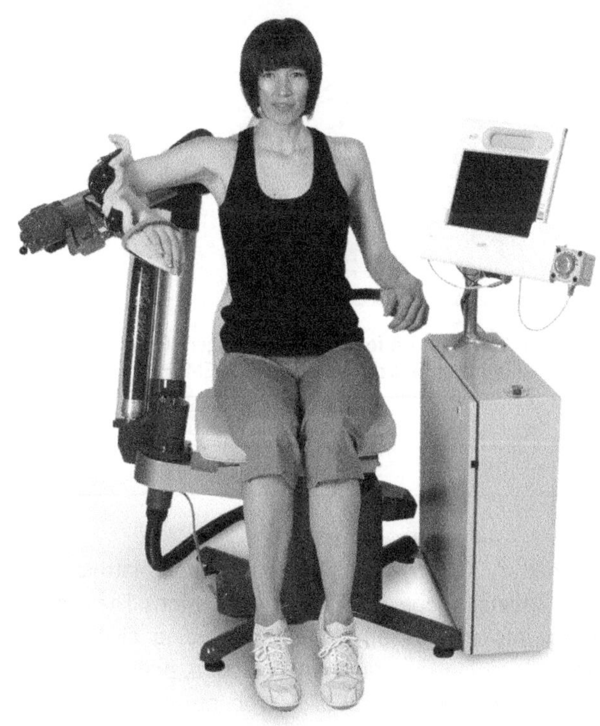

Figure B.1 CCPM device integrating user behavior autonomously.
Source: FERROBOTICS GMBH

Case 1b Active Compliant Technology

In the meantime, FerRobotics continued to improve Active Compliant Technology (ACT) and explore other markets. Through market research based on the readiness assessment, it found effective applications in the much larger automotive, aerospace, and other contact sensitive surface technology industries. Because this industrial market is much larger than the medical application markets and also has much fewer legal requirements for technology implementation, FerRobotics decided in 2011 to switch its focus to industrial markets. Today, it is the leading provider of tools for grinding polishing deburring among other tooling for ACT robots in the industrial markets and the final MRL is 10 and TRL is 9 for these markets. FerRobotics is now a world leader in the development and sale of contact intelligent and intuitive robotic equipment (Figure B.2).

Figure B.2 ACT-Based End-of-Arm Solution for Contact Intelligent Grinding of High Quality Surfaces in Automotive Industry. *Source:* FERROBOTICS GMBH

Case 2 Taurob

Company: Taurob GmbH

Taurob GmbH was founded in 2010 by Lukas Silberbauer and Matthias Biegl. Their original aim was to develop robots for firefighting missions. The company has gradually expanded into various other fields, including civil defense, CBRN*(chemical, biological, radioactive, nuclear) detection, police, academia, and, most recently, the oil & gas industry. In 2012, Taurob introduced the world's first European Union (EU) official regulations on hazardous area operations, ATEX certified (i.e., suitable for explosive environments) mobile robot.

Dr. Hasenauer, the co-author of the book, worked together with Taurob in an applied technology research project of EU to develop MTB to improve multidisciplinary communication, and develop a TRL assessment process comprising all necessary knowledge fields for functional inspection and supporting labor contents for autonomous moving design:

- Unmanned guided vehicle dedicated for hazardous missions
- Waterproof, ruggedized, and ATEX-certified
- Intuitive control system and multipurpose interfaces
- TRL focused on defining operational criteria for evaluation of perceived usefulness and perceived ease of use in case of unmanned autonomous robot action. In open air quite often robot–robot cooperation is necessary. These cooperative scenarios are men-guided
- When competitive market research showed that ATEX certification is key to becoming a strong market player, then ATEX requirements became top priority and were put into practice

MTB focused on evaluation of perceived usefulness and perceived ease of use in case of men-guided robot-to-robot open-air cooperation. MTB focused also on requirements for coal mining missions where real-time communication technology is critical.

Developing TRL assessment in compliance with requirements for coal mining missions: real-time communication technology is critical. In the context of Technology Readiness and Market Readiness, the setup of MTB and scenarios for a safety and security robot for hazardous missions in the oil and gas industry yielded a few important lessons:

a) **Cross-functionality** is essential to cope with multidisciplinary Hi-Tech projects:
   - Hi-Tech marketing inherently involves communication between multiple knowledge disciplines. Multidisciplinary communication accelerates technology acceptance by creating mutual understanding and trust in the context of safety and security.
   - Negotiations between buyers/users and suppliers/innovators
     - Require interdisciplinary cross-functional teams.
     - Both on the buyer's and the seller's side.
b) **How to overcome semantic barriers** in a multi-technology context:
   - Users/customers and suppliers/developers speak different technical languages.
   - User-environment-oriented versus technology-oriented jargon. MTB process is ideally suited to break down these barriers.
   - Initially, Hi-Tech center acted as interpreter, then each party the other's language, naïve questions, and practical examples to enable simple formulations. After that, user/supplier collaboration worked smoothly; willingness to collaborate followed soon.
c) **Behind each bottleneck lies a potential market**:
   - Highly qualified labor (e.g., firemen) is in short supply and very expensive. Statistical value of human life is high: US$ 5M–10M per qualified employee. Cost of insurance is high, tight budgets constitute bottlenecks. They open the market for robots (*Price ~$five digits*). Total cost of

ownership and lifecycle costs of robots are two orders of magnitude lower than those of humans.

Because the market need for robots in hazardous operations has been well-known, and because of the strong capabilities of Taurob for such robots, starting readiness assessment set both MRL and TRL at 2.

MTB has subsequently established USP for Taurob:

- Unmanned guided vehicle dedicated for hazardous missions.
- Waterproof, ruggedized, and ATEX-certified.
- Intuitive control system and multipurpose interfaces.

With continued market research and the USP, MRL for Taurob system went from 2 to 5, and finally 8, while TRL went from 3 to 4 and finally 9 (Figures B.3 and B.4).

Figure B.3 Taurob Tracker – 2014. *Source:* Taurob GmbH

Figure B.4 Taurob Tracker – 2023. *Source:* Taurob GmbH

Case 3 Atmospheric Plasma on Wood Surface

This case is ongoing research for a potentially disruptive innovation in the field of plasma surface activation of wood surfaces.

Experimental results indicated hydrophilic behavior and increase of adhesive forces in conformance with results of other European research groups.

To prove hydrophilic or hydrophobic behavior of the well-defined wood surface, a test ink is used. Wood surface is activated with 25 kV for three minutes and diffusion behavior of test ink is observed and measured. After many trials with different types of wood, an unexpected and not yet explained fact is that using 50 kV does not show any significant effect compared with 25 kV. As a result, it is far too early to conclude systematic confirmation, but step by step, nescience is converted into scientific results.

Moreover, activation of wood surface may result in the following research questions:

(1) Higher hydrophilic effect
(2) Higher protection against fungi
(3) Higher glue adhesiveness
(4) Substitution of primer
(5) Substitution of humid protection liquids
(6) Surface activation is also effective in tongue and groove joint

From Intrapreneurship Management viewpoint, this research was initiated by internal innovators in order to find an effective and economically and ecologically feasible solution and avoid aggressive chemicals. Research started in the transition from basic research to applied research.

The innovation shows potentially strong ecological advantages, yet, still many questions are open and with unknown answers. Therefore, readiness assessment of the project started with MRL at 1 and TRL at 2, and further exploration may provide answers to questions (1), probably (2), and hopefully (3).

Because of the exploratory nature of this project, over the years, MRL only advanced to two and TRL only to 3. However, one advantage of readiness assessment is the continued systematic documentation of assessment results which can provide insights into project development.

Case 4 Image Biopsy (IB) Lab

This case shows how nescience is transformed into knowledge in the field of finding early warning signals for structural bone disease, starting with an X-ray of a tibia bone layer after an X-ray operation. There are two major ways to analyze the information content of the X-ray:

(1) Analog mode versus digital mode: Based on the vast experience of the radiologist to recognize pathologic changes in the bone contour and in

texture of bone surface, and his vast knowledge of typical signals of early bone disease and his deep experiential knowledge in recognizing pathologic patterns, doctors' eye plays a significant role in optical search, analysis and understanding of X-ray pictures.

(2) The digital mode of analyzing X-ray photograph consists of comprehensive digital analysis of all pixels of the X-ray photograph. A pixel is the smallest addressable picture element described by a two-dimensional picture with the x and y coordinate values and a third information referring to the color of the single pixel. Granularity of digital picture is expressed by number of pixels per inch. X-ray photographs can be searched through and analyzed pixel-wise to find those pixels that show a mineral bone density lower than threshold gray value. Those pixels are nearly warning signals for systematic too low mineral bone density.

A low mineral bone density results in rough bone surface caused by loss of mineral bone material as can be seen in both pictures below. The indirect measure of roughness of bone surface can be done by the Hurst exponent or parameter, H, which is expressed by autocorrelation of surface points [1].

H lies between 0 and 1, [0<=H<=1]; the higher H, the smoother bone surface, and the higher mineral bone density. If H is significantly lower than 0.6. the risk of systematic loss of mineral bone density is greater. Executing a search by human eye is highly exhausting and monotonous and has the inherent risk of overlooking the rare single-pixel evidence of early warning signals. To reduce the risk of overlooking early warning signals, the company Image Biopsy Lab recommends using digital search procedure. Each pixel cell shows a specific gray value with varying gray darkness being a representative indicator for mineral bone density. If the gray value is below a threshold value, then this single pixel is marked (Figure B.5).

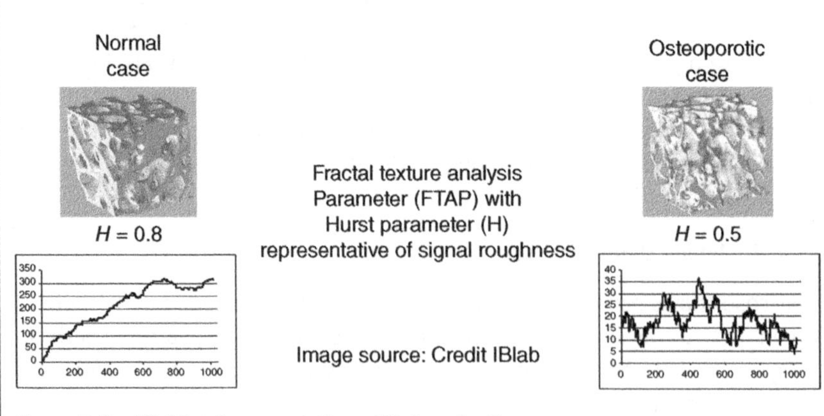

Figure B.5 3D Print Representation of Trabecular Bone

Because of the strong market need for more powerful detection and the technology still being in the research stage, for the readiness assessment, the MRL started at 2 and TRL started at 3.

Then, with a greater understanding of the market and the emerging computer solution, both MRL and TRL quickly went to 6 for both [2].

This case also shows the immense power of effective Intrapreneurship Management by an innovation team pushing forward their convincing creative idea, getting international approval from FDA, and expanding the method not only to identify structural deficits of mineral bone density but also to new applications such as measuring geometric forms as special curvature, cracks, and fissures using X-ray pixel information. Furthermore, the collective power of problem-solving and the mutually inspiring properties of the whole innovation staff with the clever division of labor and responsibility form the innovation kernel.

This IB lab technology opens several application windows that are already brought to the market; beyond bone mineral density, the technology can be used for measuring geometric data of bones as well as curvature. From pharmaceutical viewpoint, the identification of scarce early warning signals is very important for testing the effectiveness of new pharmaceutical products against early outbreaks of osteoporosis. TRL as well as MRL and the respective granted FDA certificates consolidate the competitive lead of IB lab.

The whole development process resulted in an MRL of 10 and a TRL of 9 by being granted the second FDA certificate for measuring and classifying bone geometry.

Case 5 Vanadium–Redox Flow Battery

Vanadium–Redox flow battery (VRFB), which uses vanadium ions as charge carrier with reduced oxidation, is a technology enabling and suited for stationary RAPS when coupled with photovoltaic and/or wind energy, shows low self-discharge, longevity of electrolyte electricity storage medium, and is intrinsically safe. This technology can contribute to life quality in remote areas. Despite the low energy density, the technology has been well-accepted for its high reliability and robustness.

VRFB was developed in Austria based on the similar technology used in Japan but with significantly smaller battery stacks. These smaller stacks offer a higher versatility for decentralized private use in remote areas.

Through MTB, the project developed:

- Technology acceptance and fulfillment of marketability criteria
- Volatility of raw material prices of electrolyte
- Business model variants

MTB also enabled users to generate a valid estimation of the required matching between the market challenges of the envisaged market segment and the necessary marketability criteria to market the respective innovation.

Readiness assessment yielded starting MRL as 1. The relatively low MRL is a result of low number of sales prospects. The long operational test of VRFB resulted in starting MRL of 2. The innovation project with full documentation of know-how was sold to a European system producer with further successful developments, which has resulted in MRL rising to 3 and finally 6, and TRL to 5 and finally to TRL 8 (Figure B.6).

Figure B.6 Field view of a VRFB setup. *Source:* Enerox GmbH 9.9.2023 cellcube

References

1 Soares, H.C., da Silva, L., Lobão, D.C. et al. (2013). Hurst exponent analysis of moving metallic surfaces. *Physica A* 392 (2013): 5307–5312.

2 Martinez, O.S. et al. (2013). Rough surfaces profiles and speckle patterns analysis by hurst exponent method. *Journal of Materials Science and Engineering B* 3 (12): 759–766.

Index

Intrapreneurship Management: Concepts, Methods, and Software for Managing Technological Innovation in Organizations, First Edition. Rainer Hasenauer and Oliver Yu.

Published 2024 by John Wiley & Sons, Inc.

Printed and bound by CPI Group (UK) Ltd, Croydon, CR0 4YY

16/04/2025

14658584-0001